Wetland Plants

Maryam Akram Butt • Muhammad Zafar
Mushtaq Ahmed • Shabnum Shaheen
Shazia Sultana

Wetland Plants

A Source of Nutrition and Ethno-medicines

Springer

Maryam Akram Butt
Department of Plant Sciences
Quaid-i-Azam University
Islamabad, Pakistan

Muhammad Zafar
Department of Plant Sciences
Quaid-i-Azam University
Islamabad, Pakistan

Mushtaq Ahmed
Department of Plant Sciences
Quaid-i-Azam University
Islamabad, Pakistan

Shabnum Shaheen
Department of Plant Sciences
Lahore College for Women University
Lahore, Pakistan

Shazia Sultana
Department of Plant Sciences
Quaid-i-Azam University
Islamabad, Pakistan

ISBN 978-3-030-69257-5 ISBN 978-3-030-69258-2 (eBook)
https://doi.org/10.1007/978-3-030-69258-2

This Springer imprint is published by the registered company Springer Nature Switzerland AG
The registered company address is: Gewerbestrasse 11, 6330 Cham, Switzerland

This book is dedicated to our teachers and parents

Foreword

The Himalayan region has many wetlands of different sizes and shapes, which make it a distinctive ecosystem that fulfills an essential function in the overall water cycle of the basins. Wetlands found in the Himalayan region are important as they provide shelter and food to living organisms. Pastures on the banks of wetlands are used for grazing and have also been known as the home of several rare endemic species of fauna and flora.

The biodiversity of wetlands has great economic and aesthetic value, and this diversity is helpful in maintaining overall environmental health. Most people around the world depend on aquatic resources for food, medicines, and commercial purposes such as tourism and fishing. However, due to many reasons, these wetlands are facing serious threats. Data included in this illustrated wetland guide will help readers, regardless of their botanical background, recognize the plants included in this book.

The main purpose of this book is to explore the wetland flora of the Himalayan region, to identify those plants that are beneficial to mankind, and to inform people about the challenges faced by these wetland areas and the strategies adopted for their conservation. This book not only helps with identification of wetland plants, but also helps in conservation and protection of most valuable resources.

(QAU, Islamabad) Maryam Akram Butt
(QAU, Islamabad) Muhammad Zafar
(QAU, Islamabad) Mushtaq Ahmad
(LCWU, Lahore) Shabnum Shaheen
(QAU, Islamabad) Shazia Sultana

Preface

Wetlands are a major feature of the landscape in all parts of the world, covering nearly 6% of its area (i.e., 8.6 million km^2). They are the ecotones between terrestrial and aquatic ecosystems, have unique hydrologic functions, and are extensively utilized for the sourcing of food, medicine, etc., along with shelter, thus forming dynamic and significant ecosystems needed by all living beings. Such lands include bogs, fens, marshes, peatlands, moors, swamps, bottomlands, and mangrove forest areas that may be wet year round or during certain periods of time.

Unfortunately, most wetlands and water bodies are under increasing threats as they are drying up rapidly due to various man-made impacts. Many of them are now transformed into other land forms, such as paddy fields, human settlements, and sites for developmental projects. These aquatic life forms play an important role in supplementing human diet and nutritional balance; besides, they also support the livelihood and income of a considerable section of society living around them. Unfortunately, there is little recognition of wetland landscapes for their current and potential value in supplying dietary food items. As 38% of wetlands in the country have been lost in the past 10 years and many more are under threat, there is a need to take up ethnobotanical surveys of important resources that are used locally so that an action plan can be developed to protect the ones in extensive use.

The aim of this book is to identify the most common edible wetland plant species that are either used for food or medicinal purposes or both, the nutritional values of edible wetland plants, and species that need immediate attention for conservation as per local perception as well as based on the extent of the pressure. It is expected that this study will not only document the local knowledge for the use of these plants, which may be lost in the near future as traditional cultures are eroding day by day, but also help in maintaining a linkage between local culture and its ecosystem, which is of utmost concern for the conservation of the local environments.

This volume provides complete, comprehensive, and broad subject-based reviews for students, teachers, researchers, policy makers, conservationists, and NGOs interested in the field of biodiversity and conservation of wetland plants.

Islamabad, Pakistan — Maryam Akram Butt
Muhammad Zafar
Mushtaq Ahmed
Lahore, Pakistan — Shabnum Shaheen
Islamabad, Pakistan — Shazia Sultana

Acknowledgments

We wish to express our gratefulness to all the authors who have contributed in this book. We thank them for their cooperation and erudition. We also thank several colleagues for their help in many ways and for their suggestions from time to time during the finalization of this book.

We are greatly thankful to the valuable assistance of **Dr. Mir Ajab Khan**, Ex Dean, Faculty of Biological Sciences, Quaid-i-Azam University, Islamabad, for his kindness, love, immense help, valuable guidance, keen interest, and encouragement throughout our field work. Thanks to the local people of study area for making us aware of folk uses and vernacular names of different plant species.

We wish to express our gratefulness to our respective family members for their cooperation during collection of plants. We are also thankful to all those who could not be mentioned individually, but their direct or indirect help is highly appreciated.

Contents

About the Authors

Dr. Maryam Akram Butt obtained her PhD from the Department of Plant Sciences at Quaid-I- Azam University Islamabad, Pakistan, in 2019 in the field of plant systematics and biodiversity and medicinal plants. She is a renowned scientist in the field of wetland biodiversity, medicinal plants, and plant systematics. Dr. Butt has published 20 research papers in reputed national and international journals with high IF, H & I indices. She attended has many national and international conferences and delivered lectures on wetland biodiversity. Currently, she is working as visiting faculty in the Department of Botany at the University of Kotli, Azad Jammu and Kashmir. maryamakram40@gmail.com

Professor Dr. Mushtaq Ahmad is working in the Department of Plant Sciences, Quaid-i- Azam University Islamabad, on advanced areas of plant systematics and biodiversity to identify wild plant resources as alternative sources for food, medicine, and biomass energy. Dr. Ahmad has been awarded with many research grant projects funded by GBIF, BIFA (Japan–USA), NAS-USA, PAS, HEC, and TWAS. These research grants helped Dr. Ahmad establish a modern digitized herbarium (ISL) and a nutraceutical and biofuel research laboratory. Dr. Ahmad has successfully supervised 30 PhD and 90 MPhil scholars, published 582 research papers with impact factor greater than 650, 25 international books, and 16 book chapters. Dr. Ahmad's citations have reached up to 8600 with H-index 46 and I-Index 199. He is the awardee of many international and national awards including Top 2% Influential Scientist of the World (2020); Young Research Scholar Award by HEC (2019); Highly Cited Research Paper Award by Elsevier and Willey (2019); Young Scientist Award by CAS – PIFI – China (2018); Young Membership Award by Pakistan Academy of Sciences (2016); Productive Scientist Awards by PCST (2009 to date); TTS Performance-Based Awards (2010 to date); Post Doc. Fellowship by TWAS-Malaysia (2012); Best Book Award by HEC (2013); Best Research Paper Award by HEC (2011); Gold Medal by Pakistan Academy of Sciences (2011); and membership of many international and national academic bodies (2009 to date). He has hosted a series of TV programs to raise awareness among Pakistani communities, farmers linkages with academia, and industries to use plant biodiversity for socio-economic uplifting. mushtaq@qau.edu.pk

Dr. Muhammad Zafar obtained his PhD from Quaid-I-Azam University Islamabad, Pakistan, in 2011. Currently, he is working as assistant professor in the Department of Plant Sciences, Quaid-i-Azam University Islamabad, Pakistan. He is a renowned scientist in the field of medicinal plants, plant biodiversity, and renewable energy. He is founder of the first nutraceutical and traditional medicinal plant laboratory as well as biofuel and biomass energy laboratory at Quaid-i-Azam University Islamabad, Pakistan. Dr. Zafar supervised 13 PhD and 49 MPhil research scholars and is currently supervising 5 PhD and 7 MPhil research students. He has published so far 370 research papers with high impact factor and H index in international journals of repute and above 3000 citations. Dr. Zafar has published 12 international books and chapters in the field of medicinal plant biodiversity and energy technology. The Ministry of Science and Technology and Pakistan Academy of Sciences placed Dr. Zafar in the list of 10 most influential Pakistani young scientists and awarded him with a gold medal and productive scientist award. His creative work at the university includes establishment of a medicinal plant lab, biofuel lab, digital herbarium, botanical garden, and nutraceutical lab for research purposes and linkage with industries for rural and academic development. Dr. Zafar is a member of international societies like USEA, Pakistan Botanical Society, International Society of Ethnopharmacology, and Asian Biofuels and associate editor of various journals of repute. He has attended and organized more than 100 national and international conferences and delivered plenary lectures on medicinal plants and renewable energy technologies. Currently, he is linked with world-renowned institutes for collaboration regarding research and development in natural products from medicinal plants and biomass energy technologies. zafar@qau.edu.pk

Dr. Shabnum Shaheen is a renowned scientist in the field of plant sciences. She completed her postdoc from Malaysia and PhD in plant sciences from Quaid-I Azam University Islamabad, Pakistan. She is working at Lahore College for Women University, Lahore, Pakistan. Dr. Shaheen has published around 100 research articles (to date) in national and international journals/conferences. She has also authored 10 books. One of her books is on edible wild plants as an alternative approach to food security, published in Springer in 2017. She is awardee of a gold medal at an international conference of botany. She is reviewer of many well-reputed journals and also editor of an international scientific committee. She is also a member of the World Academy of Sciences and has supervised a number of graduate, postgraduate, and PhD students.

Corresponding author: shabnum_shaheen78@hotmail.com

Dr. Shazia Sultana is a young scientist in the field of biodiversity and medicinal plants (nonedible oil seeds and herbal drugs). 205 research publications (to date) in top international journals, more than 4000 citations, with high IF, H & I indices. 10 international books published and circulated internationally. Dr. Sultana is awardee of various national and international awards. She has been selected as Young Lady from Pakistan by the Institute of Post Graduate studies, USM Malaysia, Higher Education Commission, and Pakistan Council for Science and Technology. She is also the editorial board member of several well-reputed ISI journals and book series. Currently, she is supervising and working on a number of research projects on natural plants products, new energy and low carbon technology using biofuel and nonedible oil seeds for biodiesel production, and plant biodiversity and conservation. shaziaflora@hotmail.com

List of Figures

List of Tables

Chapter 1
Wetland and Wetland Plants

1.1 What Are Wetlands?

A wetland is a diverse ecosystem of the world that is covered by water, either seasonally or permanently, where oxygen-free approaches succeed. The main feature that differentiates wetlands from other lands or water bodies is the typical flora of aquatic plants, which are adapted to the distinctive hydric soil. They are associated with many functions which includes purification of water, storage of water, processing of nutrients and carbon, maintenance of shorelines, and support to fauna and flora. They are also considering as a most diverse ecosystem on earth providing habitats to many plants and animals species. Whether any single wetland plays those features, and the way to which it does them, depends on features of that wetland and the waters and lands in their surroundings. Procedures for quickly evaluating these features, ecological health of wetland, and general condition of wetland have been established in various areas and have added to wetland protection partially by educating public through awareness programs about the functions and the services of ecosystem provided by wetlands.

Naturally wetlands occurred in every part of the world. Wetland consists of three types of water brackish, salt water and fresh water. The main types of wetlands are marsh, swamp, fen and bog; sub-types consist of flood plains, mangrove forest, vernal pool, mire, sink. Peatlands are also included as a type of wetlands. Wetlands can be non-tidal or tidal. Many wetlands are intermediate areas among aquatic ecosystems and uplands, though some are dispersed throughout the landscape in highland depressions that assemble water or in areas where ground water approaches the top surface.

The amount of water in wetlands highly depends upon the rate of precipitation on that area. Some wetlands are totally flooded while others are seasonally flooded but keep saturated soils during the drought period. When the wetlands are rarely flooded still they have ability to provide saturated soil which supports wetland-adapted

M. A. Butt et al., *Wetland Plants*, https://doi.org/10.1007/978-3-030-69258-2_1

vegetation and for hydric soil features to grow. Development of hydric soils takes place when chemical changes occurred in the soil due to the oxygen deficient environments linked with extended saturation.

Depending upon the hydrology of water different communities of plants are observed in different kinds of wetlands. The plants that grow on these wetlands are termed as hydrophytes which mean water loving plants and the produce special structures to survive in these areas. Many insects, birds, and other animal species are totally dependent on wetlands to complete their different stages of life cycle, also some species use while many other wetlands for resting, feeding or other activities.

1.2 Defining Wetlands

In the above mentioned paragraph we can describe all the features of wetland clearly but still there has been a lot of argument done to precisely explain what creates a wetland. A practical definition turn into serious in organizing habitat for legal resolutions, particularly in identifying which lands are protected by federal and state laws. Finally in 1979 the U.S. Fish and Wildlife Service defined wetlands as:

> Wetlands are lands transitional between terrestrial and aquatic systems where the water table is usually at or near the surface, or the land is covered by shallow water.

Wetlands have one or more of the following 3 characteristics:

The land supports largely hydrophytes;
The substrate is mostly un-drained hydric soil
The substrate is saturated with water or covered by shallow water at some time throughout the developing season of the year.

Keddy (2010) defined wetland as "an ecosystem that rises when accumulation by water creates soils dominated by aerobic and anaerobic processes, which in return, forces the rooted plants to adapt themselves for flooding."

1.2.1 Ramsar Convention Definition

According to Ramsar international wetland conservation wetlands are defined in two ways (Figs. 1.1, 1.2, 1.3 and 1.4):

> Wetlands are regions of fen, marsh, peatlands, whether artificial or natural, temporary or permanent, with water that is flowing or static, brackish, fresh or saline, including regions of sea water whose low tide depth does not increase from six meters.

> Wetlands may include riparian and coastal regions nearby the wetlands, and bodies of marine water or islands deeper than 6 meters at low tide lying inside the wetlands.

Fig. 1.1 Pictorial view of fresh water lakes

1.3 Origin of Wetlands

Data about the first wet area plant was found in Ordovician period (485.4 million to 443.8 million years ago), when the first land plants, which were reliant on damp substrates, started colonizing on the land. Wetland animals and the plants that

Fig. 1.2 Pictorial view of stream water

depend on them persisted to develop, and the 1st swamps and marshes created at some stages of the Devonian period (419.2 million to 358.9 million years ago). Later on swamps dominated the huge areas, such land become southern North America, during the Carboniferous period (358.9 million to 298.9 million years ago), and fragments of the Cenozoic and Mesozoic eras (252.2 million years ago to till date) were also described by the occurrence of widespread habitats of wetland.

Fig. 1.3 Pictorial view of wetland vegetation across ponds and waterfall

Wetland groups rely on availability to liquid water. During geologic historical period, the availability of water has varied according to predominant global and local climatic conditions, season, elevation, latitude and distance from each groundwater and water bodies. Due to this unevenness, the communities in wetlands from different regions of the world are the produce during different circumstances.

Fig. 1.4 Glimpses of herbaceous and tree vegetation in wetlands

During glaciers formation in the Pleistocene epoch (2.6 million to 11,700 years ago) several types of lands develop along with some wetlands. In glaciated areas, due to the ice sheets movement many landscapes came into being and the load of these ice sheets depressed the crust of earth from below. As the Pleistocene glaciers pass away throughout the Northern Hemisphere, soften carved wide, flat valleys which might be occupied nowadays through main rivers and their related floodplains and wetlands (flat land vicinity adjoining to a stream). Rough scouring of the panorama in a few areas led to low spots that full of melted snow and rainwater during mainly moist years. This procedure formed the grassland depression area of the south-central Canada and Midwest.

In a number of the coldest areas of the world, wetlands are sustained through an impermeable layer of ice that stays with in the soil at some stage in the year. This constantly ice-covered surface, or permafrost, avoids both the purification of surface water into the surface and plant interaction with groundwater mineral contents. Approximately 20–22% of whole world surface is nearby polar region or high in altitude to experience permafrost. Much of northern Eurasia and North America, in addition to this Tibetan and Mongolian Plateaus, are influenced by permafrost, and these areas host huge regions of peatlands, fens and bogs. In North America most widespread regions of the world that included bogs and fens exists In western Siberia, larch-spruce-birch forests produce a part of an large inland delta, that's the biggest connecting region of peatlands with in the world. Asian highlands in common host some of the most uncommon high-elevated wetland ecosystems.

Some wetland regions have been created in different approaches during intervals of low sea level, when water becomes locked in glaciers. In addition, the heating duration that immediately observed the maximum latest glacial episode (which ended about 11, seven hundred years ago) become marked through rivers flowing with melted ice water, hidden pieces of ice that melted and produce kettle lakes, huge lakes that is found in low regions inland, sea sidelines that moved inland, and water table of coastal areas rose with the sea. Wetlands eventually evolved alongside of lakes and margins of coastal areas, in delta regions, and throughout floodplains.

Wetlands in non-glaciated areas, including the tropics, have been evolved throughout periods characterized by barely distinct climates and therefore can be converting below present-day conditions. The peat swamp forests of Indonesia are constructed on peat up to 15 m (approximately 50 ft) thick. The accumulation of this thing took place throughout a wetter period numerous thousand years ago. Although new peat continues to be forming in locations and the area stays humid, the tropical weather of the modern-day is dry sufficient to permit the degradation of peat in a few areas.

1.4 Geographic Distribution of Wetlands

Wetlands are located all around the world in each biome, or most important life region. Some wetlands, which include tidal marshes, suit the definition of a transitional region due to the fact they arise where land and water meets.

Wetlands are maximum considerable in tropical and boreal regions, however a huge type of coastal and inland wetlands also are discovered in temperate regions. This spreading is normally because of situations that encourage an abundance of water. Wetlands also are discovered with in the warm desert biome. In temperate regions, wetlands generally are discovered close to lakes, rivers, coastlines or different places in which nearby water input go above output.

1.5 Environmental Conditions

1.5.1 Climate

Wetland formation is inspired through climatic conditions and the restrictions posed through landforms. The remaining equilibrium of evaporation and precipitation determines the timing and quantity of water obtainable for the maintenance or formation of wetland circumstances. When water flows down to the hill, and the geomorphology of the site decides where it collects as well as what subsurface layers or topography stop it from draining. Each wetland has a water signature, or hydroperiod, that is characterized through the duration, timing and amount of water within the system. Additionally, flood has biological significances, because it inhibits atmospheric oxygen from being replaced from the soil. As a result, only creatures that can bear or are adjusted to anoxic or low-oxygen or situations have a benefit in wetland atmospheres.

1.5.2 Soils

Wetlands form flooded environments stay long enough throughout the developing season to cause anaerobic areas to arise in the surface of the soil, which contains the root area. This type of soil is organic or may be derived from nutrients. Organic wetland soils, including peatland soils, comprise as a minimum 12% organic matter and are commonly acidic; additionally they possess a excessive water-retaining ability and less nutrient availability. During low oxygen concentration in soil and having low decomposition process the organic matter is formed. Minerals containing wetland soils have less than 12% organic matter, and they frequently reveal gleying, where ferric iron and manganese are reduced in the soil by anaerobic bacteria flourishing in the dwindling oxygen circumstances. The ensuing ferrous iron ($Fe2+$) will become concentrated in a deep soil deposit. In wet soils, the upper soil and top soil layers take on a gray, blue-green or black colour. Linings of pore in wetland soil containing minerals are red in color, because of plant roots, which allows pores to release oxygen into the oxygen-deficient soil. The occurrence of this oxygen in an anaerobic atmosphere dissolves some of the ferrous iron left over in the water and covered it along the lining of pore. Plants and animals waste may arise at the external layer of flooded mineral soils.

1.6 Biota

The biota of a wetland system consists of its fauna and flora. The maximum essential component affecting the biota is the period of flooding (Keddy 2010). Other essential elements consist of salinity and fertility. In fens, species are particularly depending on water chemistry. The chemistry of water flowing into wetlands determined by the source of water courses and the geological element in which it flows over (Bedford 1996) in addition to the nutrients released from organic matter in the soils and vegetation at upper altitudes in hill wetlands (Nelson et al. 2011). Biota may also vary within a wetland because of season or current flood regimes.

1.7 Wetland Flora

Contribution of wetland plants to wetland ecosystem helps to understand the key characteristics of these plants. These plants are important due to many reasons (Wiegleb 1988). Plants of wetlands form the base of food chain as they provide main source of energy to the ecosystem. These plants connect biotic factors with inorganic environment through the process of photosynthesis. These plants provide habitat to the members of other groups as; fish, epiphytic bacteria, algae, and macroinvertebrates (Carpenter and Lodge 1986; Wiegleb 1988; Cronk and Mitsch 1994). They also help in phytoremediation of water and releases oxygen in water for respiration of other creatures. Also help in binding the soil thus prevent erosion of soil from the shores and banks of water bodies.

Wetland plants, which we consider to be synonymous with wetland hydrophytes, are commonly defined as plants "growing in water or on a substrate that is at least periodically deficient in oxygen as a result of excessive water content" (Cowardin et al. 1979). This term includes both herbaceous and woody species. Tiner (2016) is credited as the first to arrange plant communities according to their hydrological preferences. Aquatic plants were defined as submerged species or those with floating leaves, while marsh plants were categorized as terrestrial plants. He further organized vegetation into various "oecological classes" based on soil conditions. Very wet soils supported two classes of plants, the hydrophytes (those in water) and the helophytes (those in marshes, i.e., emergent plants). Penfound (1952) developed a classification scheme recognizing two groups, the terrestrial plants and the hydrophytes, the latter of which included both submerged and emergent species (U.S. National Research Council 1995). Under these definitions, terrestrial species cannot tolerate flooding or soil saturation during the growing season. Aquatic species require flooding and cannot tolerate dewatering, while wetland species tolerate both (U.S. National Research Council 1995). Sculthorpe (1967) also adopted this broad definition of hydrophyte.

There are 4 main types of plants that exist in wetlands throughout the world (Natarajan et al. 2018).

- Emergents
- Submerged
- Floating
- Amphibious
  - Submerged wetland flora can develop in fresh-water and saline conditions. Some species have underwater vegetation, even as others have extensive stems to permit the vegetation to attain the surface (Hutchinson 1975). These plants act as a source of food for many wetland animals and also help in the filtration of water. Examples include eelgrass and seagrasses (Fig. 1.5).

Fig. 1.5 Pictorial view of some submerged plants

– Plants that are rooted in the water bottom but their aerial parts extend out of water are termed as emergent plants. Usually they are found growing in the banks of fresh water reservoirs where water is less deep (Fig. 1.6).
– Floating plants floats on the surface of water. They are dividing into three types: Free floating, submersed floating and trailing floating. Free floating plants freely floats on the surface of water; submersed floating plants attach themselves with roots from bottom of the water while trailing plants are rooted into the bank of water bodies and have a creeping growth over the water surface (Fig. 1.7).

Fig. 1.6 Pictorial view of some emergents

Fig. 1.7 Pictorial view of floating plants

- Amphibious plants live both in wetlands and on land they produce special structures to survive in both environments (Fig. 1.8).

1.8 Plants that Grow in Land & Water

Just as amphibian animals can stay in water or on land, there also are species of plants which can continue to exist in each environment. These specific trees and plants have adapted to residing in each environment and may be observed in

Fig. 1.8 Pictorial view of amphibious plants

numerous climates. They are of different sizes starting from bald cypress tree, to very small plants like the American sponge plant.

1.8.1 Mangroves

There are 55 distinct species of mangroves that may be observed in tropical and subtropical climates across the world. They develop at the banks of streams, rivers and lakes and may live on in areas in which the temperature would not fall under

70 °F all through the winter. Different species of mangroves may be observed close to clean and salty water. The colour of the flowers and bark purely depend on the species.

1.8.2 Salix Trees

These are the trees that grow in rivers and streams, as well as in swampy areas. They are very important part of ecology of their local habitat and are also used in making household items. They also provide shelter to many wetland animals and birds.

1.9 Difference Between Terrestrial and Wetland Plants

There is a great difference among terrestrial and wetland flora, however the lines among them are often not cleared. This is due to the fact there are numerous flora which can be technically taken into consideration to be terrestrial flora due to the fact they develop in or on land however are capable of tolerate submersion in water and might even thrive in environments in which they may be exposed frequently to water. Many species of plants have unique differences in their very own species, and some of those can manage submersion in water, while different species from the identical plant family perish in those conditions.

In environments in which heavy precipitation or flooding is common, flora that may live on both dry and moist situations overlap the road among terrestrial and water bodies. There are terrestrial floras that may live on durations of waterlogging or heavy rain and still continue to exist; however, those would not be taken into consideration wetland flora. Many species of terrestrial flora can tolerate inundation or submersion of water for a quick period; however extended exposure often results in death.

Land vegetation is normally taller, wider and have well branched root system. Water plants are naturally having spongier leaves and wider which allows the plant to freely float on water body. Roots of these plants are either anchored in water bottom or floats freely on the surface of water.

1.10 Similarities Between Land and Wetland Plants

Apart from differences both types of plants has same characteristics also. Both diversities of the plants are green in colour, having leaves, well developed roots and have capacity to produce flower. Both types of plants prepare their food by the process of photosynthesis. Both perform the transpiration process. They both require blue wavelength of light to complete their process of photosynthesis. Additionally land plants also use red light for making food.

References

Bedford BL (1996) The need to define hydrologic equivalence at the landscape scale for freshwater wetland mitigation. Ecol Appl 6(1):57–68

Carpenter SR, Lodge DM (1986) Effects of submersed macrophytes on ecosystem processes. Aquat Bot 26:341–370

Cowardin LM, Carter V, Golet FC, LaRoe ET (1979) Classification of wetlands and deepwater habitats of the United States. US Department of the Interior, Fish and Wildlife Service, Washington, DC

Cronk JK, Mitsch WJ (1994) Aquatic metabolism in four newly constructed freshwater wetlands with different hydrologic inputs. Ecol Eng 3(4):449–468

Hutchinson GE (1975) A treatise on limnology: limnological botany, vol 3. Wiley, New York

Keddy PA (2010) Wetland ecology: principles and conservation. Cambridge University Press, Cambridge

Natarajan SK, Hagare D, Maheshwari B (2018) Constructed wetlands for improving stormwater quality and health of urban lakes. Water Sci Technol Water Supply 18(3):956–967

National Research Council (US). Committee on Shipborne Wastes (1995) Clean ships, clean ports, clean oceans: controlling garbage and plastic wastes at sea. National Academies Press, Washington, DC

Nelson ML, Rhoades CC, Dwire KA (2011) Influence of bedrock geology on water chemistry of slope wetlands and headwater streams in the southern Rocky Mountains. Wetlands 31(2):251–261

Penfound WT (1952) Southern swamps and marshes. Bot Rev 18(6):413–446

Sculthorpe CD (1967). Biology of vascular plants. 610 S. London. Edward Arnold Ltd.

Tiner RW (2016) Wetland indicators: a guide to wetland formation, identification, delineation, classification, and mapping. CRC Press, Boca Raton

Wiegleb G (1988) Notes on pondweeds-outlines for a monographical treatment of the genus Potamogeton L. Feddes Repert 99:249–266

Chapter 2
Climate of Wetlands

2.1 Introduction

6% of the world's area is covered by wetlands which have nearly 12% of the world's carbon pond, playing a vital part in the overall world carbon cycle (International Panel on Climate Change (IPCC) 1996; Sahagian and Melack 1998; Ferrati et al. 2005). Wetlands are taken into consideration on one of the major unknowns of the nearby future about matter fluxes and element dynamics (IPCC 2001; Paul et al. 2006). Nevertheless, recovery practitioners have to take change in climate under consideration while enforcing recovery initiatives and policymakers need to promote wetland recovery as a part of a weather change variation and mitigation strategies.

2.2 Climate Change and Wetlands

Change in climate is identified as a main risk to the existence of species and reliability of ecosystems throughout the world (Hulme 2005). The form of works on the hydrological and ecological effects projected to result from climate change has developed significantly over the previous period.

Pressures on wetlands are probably to be mediated via modifications in hydrology, oblique and direct results of modifications in temperatures, in addition to land use variation (Ferrati et al. 2005). Examples of affects because of projected modifications in intense climate events (Ramsar 2004) contain: change in base flows; altered hydrology; increased heat stress in wildlife; extended range and activity of some pest and disease vectors; increased flooding, landslide, avalanche, and mudslide damage; increased soil erosion; increased flood runoff resulting in a decrease in recharge of some floodplain aquifers; decreased water resource quantity and

M. A. Butt et al., *Wetland Plants*, https://doi.org/10.1007/978-3-030-69258-2_2

quality; increased risk of fires; increased coastal erosion and damage to coastal buildings and infrastructure; increased damage to coastal ecosystems such as coral reefs and mangroves and increased tropical cyclone activity. Under currently predicted future climate scenarios, the spread of exotics will probably be enhanced, which could increase pressure on watersheds and ecosystems (Root et al. 2003).

Climate alternate may be anticipated to act along with a number of different pressures, a lot of which, relying at the region, may also pose a ways more instant challenge for wetlands and their water sources within the short to medium term. Wetland structures are susceptible and specifically vulnerable to modifications in quality and quantity of water supply. It seems that variation in climate may also have its maximum reported effect on wetlands through modifications in hydrological regimes: specifically, the variability and nature of the hydro-period and the wide severity and variety of intense events. However, different variables associated with weather may also play vital roles in figuring out local and nearby impacts, which includes increased in temperature & altered evapo-transpiration, changed bio-geochemistry, different patterns and quantities of suspended loadings of sediment, oxidation of natural sediments, fire and the physical effects of wave energy (Watson et al. 1998; Burkett and Kusler 2000).

Change in climate will have an effect on the hydrology of individual wetland ecosystems mainly by modifications in temperature and rainfall regimes with exceptional worldwide variability. From the angle of evaluation of weather variability and the impact on wetlands, those ecosystems want to be considered with inside the broader context in their spatial area in a watershed inside a particular place. Given the range of wetland sorts and their person characteristics, the influences because of weather extrade can be fairly custom designed and so will the recuperation remedies. It can be severely critical to decide unique anticipated destiny adjustments in weather through place and behavior good enough tracking to check how real situations song with the unique weather extrade version for a place. This may also show to be tough and could take a tremendous instructional attempt to persuade governments and agencies to invest in tracking.

There is plentiful literature predicting the outcomes of weather extrade on species' ranges, however weather extrade models are hardly ever included into recovery and conservation plans. The spatial and temporal scales of weather fashions are disconnected from the scales of land parcels and moves that managers need to work within. Some studies outcomes reveal that excessive drought can purpose surprising and dramatic modifications withinside the abundance and spatial association of dominant vegetation, and that site traits will differentially have an effect on the dominant species that symbolize many flora types. They advise that the important thing to preserving resilient populations of dominant vegetation can be to preserve regions which might be situation to a huge kind of environmental extremes, together with sites which might be below pressure, even as restoring habitat structure to raise uncommon habitat abundance and decrease water pressure on dominant plant populations (Gitlin et al. 2006).

A variety of case research currently undertaken through The Wildlife Society discovered the complexity and capability results of weather change, even as additionally demonstrating the uncertainty. For example, one case take a look at cautioned that waterfowl could be vulnerable to modifications in precipitation and temperature, each of which have an effect on shallow seasonal wetlands with which the species are associated. The results will differentiate through species or even inside species based upon geographic location. The annual migration of neotropical migrant birds exposes them to weather modifications in both their breeding habitats and wintering, in addition to in migration corridors. The breeding variety of many species is carefully tied to climatic conditions, suggesting widespread breeding variety shifts are possibly as weather keeps modifying. The negative results of weather extrade on flora and fauna and their habitats can be minimized or avoided in a few instances via control movements initiated now (Inkley et al. 2004). To do so, we have to recognize the character of climatic and ecological modifications which are likely to arise locally so as to properly design wetland control and recovery plans.

2.3 Wetland Habitat Responses to Climate Change

Changes occurred in climatic conditions will most probably effect wetland habitats in a different way on a local and mega-watershed level; hence it is essential to identify that particular restoration and management problems would require exam through habitat. A mega-watershed is a landscape created from more than one watershed (Fig. 2.1).

2.3.1 Floodplains

Flood plain is a wide word used to mention one or more types of wetlands. Examples of flood plains are seasonally flooded grasslands (containing natural wet meadows), woodlands, shrub lands and tree forests (Ramsar Classification System for Wetland Type 1971). Riverine floodplains cover an area of 29,106 km^2; however, they're some of the maximum biologically frequent and vulnerable ecosystems because of the prevalence of levee systems, dams and additional changes to rivers, all of which makes them great contenders for recovery.

The degradation of floodplains is intently related to the fast decrease in biodiversity of freshwater; the major cause for the decrease in biodiversity of rivers is flow habitat alteration, and control on floods, invasion of species and huge amount of water pollution. In many countries most of the flood plains are now converted into cultivated lands therefore flood plains extinct. In emerging countries, the lingering natural flood plains are vanishing at an increase speed rate, mainly due to change in hydrology (Fig. 2.2).

Floodplains have certainly fascinated human settlement because of their nearness to water supplies, flat landscape and fertile soils which lead them to appealing

Fig. 2.1 Some examples of wetland habitat types

for constructing and building. As a result, a massive percentage of the worldwide populace now lives on river floodplains. However, a lot of those areas face flooding risks.

There is a dire need to conserve remaining, whole floodplain waterways as planned worldwide sources and to initiate to repair sediment transport hydrologic dynamics, and wetland flora to those watercourses that maintain some level of ecological reliability. Or else, intense disappearances of riparian, aquatic species and ecosystem facilities are met within few years (Tockner and Stanford 2002).

Fig. 2.2 Pictorial view of flood plains

2.3.2 *Mangroves/Intertidal Forested Wetlands*

Many of coastal and wetland ecosystems are affected by the variation in climatic conditions (Poff and Hart 2002). Generally, it is the quality of mangroves that they respond faster due to slight change in climate and sea level due to inland or seaward movement (Parkinson et al. 1994) facilitated by local geography of soil (Bacon 1994), if huge changes occur in sea-level than whole ecosystem of mangroves will collapse (Ellison 1993). In the future, inland movement of mangrove species will likely be restricted equally by coastal development or in situ differences in growth and by related anthropogenic obstacles (Ellison and Farnsworth 1996). As with diverse wetland species, interspecific variant in physiological responses of various mangrove species to elements linked with climatic variations might be anticipated to cause modifications in species arrangement and community structure following expected modifications in sea level and atmospheric carbondioxide levels (Ellison and Farnsworth 1997).

In the fast term, shielding and restoring extensive quantities of mangrove habitat is critical to mitigate a few weather change influences along with attenuating expanded incidences of floods and catastrophic tropical cyclones. In the lengthy term, notion have to receive to setting up new zones of mangrove habitat wherein there may be no struggle with human improvement in order that as sea degree rises and mangroves die-returned they're replenishing themselves on the landward volume of the intertidal zone. If this isn't always done, with in the future, the widespread regions of mangrove woodland may be long gone and with it the large engine that offers the carbon base of the tropical marine ecosystem.

Tri et al. (1998) enumerated that many cost-effective profits of mangrove recovery tied to sea protection structures in 3 coastal districts in northern Vietnam. The outcomes from the financial version display that mangrove recovery is appropriate from a profitable viewpoint created only on the uninterrupted profits of usage by local people. The recovery circumstances have even greater cost-benefit percentages while the oblique advantages of the eluded conservation rate of the sea dike system, secure rom coastal hurricane surges through the mangroves, are involved. A tough case for mangrove restoration may be made as an essential factor of a maintainable coastal control strategy (Tri et al. 1998). The association amongst those coastal populations not sheltered by mangroves and the causing huge damage of property and life as a result of the 2005 tsunami turned into major, and has caused accelerating the recovery of mangrove habitat alongside a few Indian Ocean shorelines.

2.3.3 *Marine Subtidal Aquatic Beds*

The long-time period ability of seagrasses to exist, specifically with in the tropics and subtropics, depend upon their capacity to adjust to changes in salinity managements encouraged through anthropogenic changes of upstream water, as well as projected long-term increase in temperature (Short and Neckles 1999). Tropical species are dwelling at the threshold in their higher physiological bounds of temperature (Walker et al. 1988) and salinity (Koch et al. 2007), so additional rises in salinity due to climate change and river water removal may have important values for tropical seagrasses mainly in estuaries with limited flow and great amount of loss (Koch et al. 2007). In different regions huge rainfall may also rise freshwater runoff and decrease salinity degrees inflicting decrease in cover of seagrass. Seagrass conservation has been carried out at numerous scales for more than 30 years with restrained success. Just like mangroves, they might be "held out" of life in few coastal areas as they are under constant pressure of human actions which include contamination have decreased the resiliency of these territories.

2.3.4 Salt Marshes

Change in climatic conditions can directly affect salt marshes in many ways, which includes rise of sea-level, especially while sea partitions prevent marsh flora from shifting inland and upward. While it is evident in some areas that due to rise of sea water does not means that marshes are lost some marshes may accumulate perpendicularly and keep their altitude with respect to the level of sea where sediment supply is enough. However, organogenic marsh areas and sediment areas may be more restrictive and more vulnerable to coastal squeeze, as many different marshes, if a few excessive predictions of enhanced amounts of sea level increase are recognized (Hughes 2004). McKee et al. (2004) recommends that decrease in precipitation and increase in temperature is linked with change in climate which may intensely move the tidal marshes. Rise in temperature may relate with further stressors to harm coastal marshes (Fig. 2.3).

A critical end result of growing temperature alongside the northern Gulf of Mexico will probably be a northward migration of mangroves changing salt marshes. Mangroves are tropical coastal forests which might be freeze intolerant. Chen and Twilley (1998) reported a version of mangrove reaction to freeze frequency. They witnessed that once freezes took place once in 8 years, death of mangrove forests occurred. Freeze rate of 12 years, mangroves substituted by salt

Fig. 2.3 Salt marshes

marshes. Along the Louisiana coast, freezes usually took place approximately each 4 years. In 2004 spring, though, an extreme freeze had not happened for 15 years and small mangroves arise over a big place close to the coast. If this development lingers, mangroves will possibly extent over most area of northern Gulf and some part of the south Atlantic coast. In fact, mangroves are previously becoming well-known and extensively spread due to increase in temperature (Day Jr et al. 2005).

Salt marshes also change the climatic effect. In some countries Salt marshes are most active ecosystems as it slows the global warming. These marshes trapped organic matter to keep massive quantity of carbon out of the atmosphere, decompose it, and hiding it quite rapidly.

As compared to tropical rainforest they bury organic material about 55 times faster. They even occupy less land as compared to tropical land forests; salt marshes are being able to seize more carbon.

2.3.5 Arctic Wetlands/Tundra Wetlands

Models of climate usually approve that the finest warming because of the improved greenhouse impact can also additionally arise at northern excessive latitudes and especially within the wintry weather. In addition, the rainfall at high latitude areas is mostly projected to raise, both in winter and in summer (Houghton 2001). For water sources, all weather situations lead (with excessive confidence) to the huge-scale lack of snowpack at mild elevations through mid-century, bringing massive reductions in summer time season flow in all rivers and streams that rely on melting of snow (Mote et al. 2003). Where consistent supply of water is accessible throughout the warm season that exceeds the need of vaporization and outflow losses, the soil stays saturated and excessive water table is conserved (Woo and Young 2006). Though, a constant warming tendency below climate change will eradicate these remaining snow banks. So many melted water wetlands will disappeared. While the mutual influence of rainfall and higher temperature is still unclear, it appears possibly that snow blocks in those regions will decline, and evapo-transpiration will rise (Dankers and Christensen 2005). These phenomena would require vast modifications to be made within the control and recovery of wetlands on this region.

For widespread wetlands, variation in the equilibrium of water to improve the process of evaporation (because of longer and hotter summer period than the present) will not only cause loss of water from the patches of wetland, but it also decrease the inputs of water from their areas. Hence, it causes serious damage to many wetland patches. Greater melting of glaciers due to increase in climatic conditions might also decrease the water table, that adversely affects most surviving wetlands, increase in activities of thermokarst leads to flooding that can create new wetlands, or to switch from fens to bogs.

2.3.6 *Forested and Non-Forested Peatlands*

Peatlands are significant natural ecosystems with great importance for conservation of biodiversity, for regulation of climate and for the welfare of human. Peatlands are formed by the accumulation of peat (organic matter) resulting from decomposed and dead plant material which are under permanent water saturation. Throughout the world it covers approx. four million km^2, occurs in almost 180 countries and denotes at least a 3rd of the world's wetland source (Parish et al. 2008).

Dynamics of peatland are very subtle to changes in the water cycle, which return to distinctions in the carbon cycle and climate (Briggs et al. 2007). The reaction of peatlands to alternate with in the climatic water budget is important to predicting ability feedbacks on the worldwide carbon cycle (Belyea and Malmer 2004). Changes in peatland environment capabilities can be mediated via land-use changes or climatic warming. In each case, decreasing of the water degree can be the main factor. Logically, decreased water ranges with the resultant growth in oxygen availability with in the upper soil can be assumed to bring about increased costs of peat decomposition (Laiho 2006). Due to climatic condition the melting of glaciers and desertification of steppe peatlands occurs. In the future, influences of weather change on peatlands are anticipated to seriously increase. Tropical, mountain and coastal peatlands are all anticipated to be in particularly vulnerable (Parish et al. 2008). There are numerous gaps in our understanding of the carbon cycle in peatlands below change, such as: how the quantities and pleasant parameters of clutter inputs alternate in unique peatland sites after short- and lengthy-time period alternate in the water stage; and the way the litters produced with the aid of using the successional plant life groups decompose below the modified environmental situations following continual decreasing of the water stage in the lengthy time period (Laiho 2006). Restoring and protecting peatlands is also critical to conserving the hydrological functions and biodiversity they provide (Fig. 2.4).

2.3.7 *Freshwater Marshes and Tree-Dominated Wetlands*

These classifications incorporate a vast variety of habitat types, with high-quality degrees of hydro-period and intensity of inundation, which includes vernal swimming pools and moist prairies with a moist season water table at or slightly above the floor for a totally short duration, to hardwood swamps, cypress swamps, bulrush marshes and sawgrass inundated by through almost a meter of water for few months. Most of those habitats reply specially to mild modifications in water quality and hydrology. There is a great wealth of literature stemming from many a long time of studies at the capabilities and control of those wetland systems, which includes restoration, however now no longer weather change (Fig. 2.5).

Based at the synergistic impact of more than one stressors, the control and recovery of those habitats can be extra hard with in the future because of the existing

Fig. 2.4 Non forested peatlands view

Fig. 2.5 Tree dominated wetland view

availability of many extra efficient colonizer species including *Lygodium microphyllum*, *Phragmites* and *Imperata cylindrica*. Given the original reactions of the many endemic species held by these habitations, a huge variety of diffused environmental modifications may want to decrease their sustainability and increase the danger of species extinction. These elements will need to be taken into consideration to study new rules and suggestions for wetland control and recovery.

2.4 Impact of Climate Change to Rivers, Lakes and Streams

Changes in climatic conditions like increase in temperatures, severe storms and variations in seasonal rate of rainfall will affect rivers, streams and lakes. Water temperature is raised with increasing atmospheric temperature. Water surface is mainly affected by these changes, which declines the obtainability of fresh water habitation for species which lives in cold water. In deep lakes increase in water temperature slower down all life processes which helps in adding oxygen to the water resulting in the formation of dead zones. In these zones harmful algal blooms occurred and also large scale of fish mortality occurs.

Increase in temperature causes melting of snow and glaciers that joined with severe rainfall cause flooding which will affect the reproduction capacities of aquatic species. Aquatic plant and animal species are in great stress due to variations in the timing of low and high watercourse flow. Flooding on regular basis leads to outbreaks of water-borne bacteria which result in poor quality water which is injurious to health (Fig. 2.6).

2.5 Impact of Climate Change to Water Sources

The change in climatic conditions threatens the quality of water sources by complete runoff of sediment and pollutants, reduced water accessibility from saltwater intrusion and drought, as well as poorly distressing whole efforts to sustain quality of water (Fig. 2.7).

Due to change in climate heavy rainstorms are projected to increase, which increases runoff of pollutants and sediments in water sources such as lakes, rivers and streams which result in complicate management of drinking water services and hence increased in costs takes place. Increased soil erosion & sedimentation can reduce the quality of water, block the storm water managing systems and decline in storing ability.

In many countries throughout the world due to sudden change in climate the drought condition occurred. In drought condition the water levels in soil faces a huge loss and on the other hand demand of water increases. In other way we can say that increase in temperature causes either drought or increase in sea level.

Quality of water is highly affected by the change in climatic conditions. Excess water from storms can damage quality of water and aggravate remaining issue of pollution. Due to increase in atmospheric temperature the water temperature automatically increased which can promote the growth of water microbes and algae in waterbodies. Due to presence of these microbes and algae the water become polluted and not taken as drinking purposes.

Fig. 2.6 Increase in river flow due to melting of glaciers

2.6 Impact of Climate Change to Quality of Water

Change in climatic condition can cause change in patterns of rainfall, atmospheric temperature and temperature of associated water, it also increases amounts of sedimentation and soil erosion. These modifications might cause poor quality water by increasing storm water runoff, more soil erosion and sedimentation, threaten water sources, and also harmful algal blooms produces in high amount. Modifications in quality of water from fresh water sources may also reduce the quality of obtainable source.

Fig. 2.7 Spring of drinking water

2.6.1 Storm Water Runoff

Estimated rises in the intensity and frequency of huge rainfall can produce excess of storm water which can transport injurious sediments, pollutants and nutrients, into waterbodies.

Storm water runoff is precipitation that runs over the surface of ground. It is formed while rain showers on driveways, roads, rooftops parking areas and different concreted planes that don't permit water to infuse into them. In urban areas the runoff of storm water is the major cause of watercourse diminishing. In paved surfaces when precipitation occurred the amount of runoff is greater as compared to the forest areas where runoff is low. These massive volumes of water are rapidly carried to nearby lakes, streams, rivers and wetlands and might cause soil erosion and floods, and wash away vital habitat for creatures that stay inside these water bodies.

2.6.2 Soil Erosion and Sedimentation

Huge rainstorms and more common and extreme hurricanes can transport residue in storm water in excess amount to increase stream and river speed which result in higher rate of soil erosion (Fig. 2.8).

Carbon is found in huge amount in the soil in the form of organic matter. This carbon is good for the growth and development of plants. This stock of carbon in soil is effected by human activities excessive cutting of forests, increased tillage and urban land formation, and agricultural and forestry practices. Due to these activities the breakdown of organic matter occurs which in result convert carbon to carbondioxide which is the main source of global warming, which is excreted from the soil.

Fig. 2.8 Erosion of soil due to climatic changes

2.6.3 Algal Blooms

Increase in atmospheric temperatures may have a equivalent influence on temperature of water. Hot waters, mixed increased concentration of nutrients, can result in greater outbursts of unsafe blooms of algae (Fig. 2.9).

Fig. 2.9 Harmful algal blooms in streams and ponds due to climatic change

2.7 Climate Implications – Drinking Water Quality

The quality and the quantity of drinking water may be affected by the climatic implications. Change in climate may increase intensity and frequency of storms; increases air temperatures and alter drought conditions.

Water sources are essential to each ecosystem and society. All humans rely on a fresh source of drinking water to maintain their health. They also require water for energy production, agriculture, recreation, navigation and industrial products manufacturing. All of these functions put an extra pressure on water sources and are stresses which might be intensified by climatic change. In few regions, shortage of water is not a problem but the major problem is increase in water level in rivers, streams which result in rise in sea level. Due to this the water quality is affected badly and this also destroys the structures that are used to deliver and transport water.

Huge rainstorms can upturn the amount of excess water into lakes and rivers, nutrients, washing sediment, trash, pollutants, livestock wastes and other things into water supplies, making them useless, risky or requirement of water treatment. More common floods and storms result in runoffs from treatment plants and sewage wastes into river, lakes and streams and recycled for drinking water.

Fig. 2.10 Flooded water due to heavy rainfall

2.7.1 Health Impacts

Overflows from sewage structures and therapeutic plants may also result in an increase in the dominance of water-borne microbes, like *Giardia* and *Cryptosporidium*. Health damage due to these water-borne microbes could be severe.

Due to change in climate it directly affects the ground and surface water and also the water resources. Drinking this water can cause some serious health issues including gastrointestinal infection such as dysentery, diarrhea, damage the respiratory and nervous systems of body, or kidney and liver disorders (Fig. 2.10).

References

Bacon PR (1994) Template for evaluation of impacts of sea level rise on Caribbean coastal wetlands. Ecol Eng 3(2):171–186

Belyea LR, Malmer N (2004) Carbon sequestration in peatland: patterns and mechanisms of response to climate change. Glob Chang Biol 10(7):1043–1052

Briggs J, Large DJ, Snape C, Drage T, Whittles D, Cooper M, Macquaker JH, Spiro BF (2007) Influence of climate and hydrology on carbon in an early Miocene peatland. Earth Planet Sci Lett 253(3–4):445–454

Burkett V, Kusler J (2000) Climate change: potential impacts and interactions in wetlands of the United States 1. J Am Water Resour Assoc 36(2):313–320

Chen R, Twilley RR (1998) A gap dynamic model of mangrove forest development along gradients of soil salinity and nutrient resources. J Ecol 86(1):37–51

Dankers R, Christensen OB (2005) Climate change impact on snow coverage, evaporation and river discharge in the sub-arctic Tana Basin, northern Fennoscandia. Clim Chang 69(2–3):367–392

Day JW Jr, Barras J, Clairain E, Johnston J, Justic D, Kemp GP, Ko JY, Lane R, Mitsch WJ, Steyer G, Templet P (2005) Implications of global climatic change and energy cost and availability for the restoration of the Mississippi delta. Ecol Eng 24(4):253–265

Ellison JC (1993) Mangrove retreat with rising sea-level, Bermuda. Estuar Coast Shelf Sci 37(1):75–87

Ellison AM, Farnsworth EJ (1996) Anthropogenic disturbance of Caribbean mangrove ecosystems: past impacts, present trends, and future predictions. Biotropica 28:549–565

Ellison AM, Farnsworth EJ (1997) Simulated sea level change alters anatomy, physiology, growth, and reproduction of red mangrove (Rhizophora mangle L.). Oecologia 112(4):435–446

Ferrati R, Canziani GA, Moreno DR (2005) Esteros del Ibera: hydrometeorological and hydrological characterization. Ecol Model 186(1):3–15

Gitlin AR, Sthultz CM, Bowker MA, Stumpf S, Paxton KL, Kennedy K, Muñoz A, Bailey JK, Whitham TG (2006) Mortality gradients within and among dominant plant populations as barometers of ecosystem change during extreme drought. Conserv Biol 20(5):1477–1486

Houghton JT (2001) Climate change 2001: the scientific basis. Cambridge University Press, Cambridge

Hughes RG (2004) Climate change and loss of saltmarshes: consequences for birds. Ibis 146:21–28

Hulme PE (2005) Adapting to climate change: is there scope for ecological management in the face of a global threat? J Appl Ecol 42(5):784–794
Inkley DB, Anderson MG, Blaustein AR, Burkett V, Felzer B, Griffith B, Price J, Root TL (2004) Global climate change and wildlife in North America, vol 04-2. Wildlife Society, Bethesda, pp 1–26
IPCC (Intergovernmental Panel on Climate Change) (1996) Climate change 1995: the science of climate change, The second assessment report of the IPCC: contribution of Working Group I, 572. Cambridge University Press, New York
IPCC (Intergovernmental Panel on Climate Change) (2001) Climate change 2001: the scientific basis, Contribution of working group I to the third assessment report of the Intergovernmental Panel on Climate Change. Cambridge University Press, Cambridge
Koch MS, Schopmeyer SA, Kyhn-Hansen C, Madden CJ, Peters JS (2007) Tropical seagrass species tolerance to hypersalinity stress. Aquat Bot 86(1):14–24
Laiho R (2006) Decomposition in peatlands: reconciling seemingly contrasting results on the impacts of lowered water levels. Soil Biol Biochem 38(8):2011–2024
McKee KL, Mendelssohn IA, Materne D, M. (2004) Acute salt marsh dieback in the Mississippi River deltaic plain: a drought-induced phenomenon? Glob Ecol Biogeogr 13(1):65–73
Mote PW, Parson EA, Hamlet AF, Keeton WS, Lettenmaier D, Mantua N, Miles EL, Peterson DW, Peterson DL, Slaughter R, Snover AK (2003) Preparing for climatic change: the water, salmon, and forests of the Pacific northwest. Clim Chang 61(1–2):45–88
Parish F, Sirin A, Charman D, Joosten H, Minayeva T, Silvius M, Stringer L (eds) (2008) Assessment on peatlands, biodiversity and climate change: main report. Global Environment Centre, Kuala Lumpur and Wetlands International, Wageningen 179 pp
Parkinson RW, DeLaune RD, White JR (1994) Holocene sea-level rise and the fate of mangrove forests within the wider Caribbean region. J Coast Res 10:1077–1086
Paul S, Küsel K, Alewell C (2006) Reduction processes in forest wetlands: tracking down heterogeneity of source/sink functions with a combination of methods. Soil Biol Biochem 38(5):1028–1039
Poff NL, Hart DD (2002) How dams vary and why it matters for the emerging science of dam removal: an ecological classification of dams is needed to characterize how the tremendous variation in the size, operational mode, age, and number of dams in a river basin influences the potential for restoring regulated rivers via dam removal. Bioscience 52(8):659–668
Ramsar C (2004) The list of wetlands of international importance. RAMSAR Secretariat, Gland
Root TL, Price JT, Hall KR, Schneider SH, Rosenzweig C, Pounds JA (2003) Fingerprints of global warming on wild animals and plants. Nature 421(6918):57–60
Sahagian D, Melack J (1998) Global wetland distribution and functional characterization: trace gases and the hydrologic cycle. Report from the joint GAIM. IGBP-DIS/IGAC/LUCC workshop, IGBP report. IGBP Secretariat, Stockholm
Short FT, Neckles HA (1999) The effects of global climate change on seagrasses. Aquat Bot 63(3–4):169–196
Tockner K, Stanford JA (2002) Riverine flood plains: present state and future trends. Environ Conserv 29:308–330
Tri NH, Adger WN, Kelly PM (1998) Natural resource management in mitigating climate impacts: the example of mangrove restoration in Vietnam. Glob Environ Chang 8(1):49–61
Walker DI, Kendrick GA, McComb AJ (1988) The distribution of seagrass species in Shark Bay, Western Australia, with notes on their ecology. Aquat Bot 30(4):305–317
Watson RT, Zinyowera MC, Moss RH, Dokken DJ (1998) The regional impacts of climate change. IPCC, Geneva
Woo MK, Young KL (2006) High Arctic wetlands: their occurrence, hydrological characteristics and sustainability. J Hydrol 320(3–4):432–450

Chapter 3
Types of Wetland and Wetland Plants

3.1 Introduction

Wetlands act as a boundary between water and land. Wetlands are comprised of three factors i.e. hydrologic periods, hydric soils, and having water-loving plants. Usually, in wetlands, water saturation is the leading factor that determines the nature of the soil and the types of fauna and flora living on the surface of the soil or inside the soil. Wetlands usually vary because of local and regional dissimilarities in topography, soils, hydrology, climate, vegetation, water chemistry, and other causes, which also includes disturbance by humans.

For monitoring purposes under the Clean Water Act, the word wetlands mean "those regions that are saturated or inundated by ground or surface water at a rate and interval appropriate to support", and that below ordinary situations do support, an occurrence of flora usually adapted for existence in water-logged soil surroundings. There are 3 main functions of wetlands.

- It provides habitat for animals and plants that use wetlands as their homes.
- It also controls flood conditions. They can store large quantities of water during heavy rains.
- They purify the water from injurious chemicals like herbicides, pesticides, pathogens, and pollutants.

3.2 Wetlands

A wetland is a region where the soil is covered with water, or it is present either near the soil surface or at the soil throughout the year or sometimes for varying periods throughout the year, along with the growing season. Water in large part determines the developmental stages of soil and the forms of animals and plant populations

M. A. Butt et al., *Wetland Plants*, https://doi.org/10.1007/978-3-030-69258-2_3

existing in and on the soil. It also supports both terrestrial and aquatic plant and animal species. Long term presence of water generates circumstances that promote the growth of plants and stimulates the soil development in the wetland zone. Wetlands differ commonly due to local and regional changes in topography, soil, water, climate, flora, and other factors.

It has been commonly observed that all wetlands are wet, but some wetlands are seasonal and depend on rainfall. Wetlands are transition areas where the cycling of nutrients, the energy of the sun, and the flow of water produce a specific ecosystem that is categorized by soils, vegetation, and hydrology creating these zones very essential features of a watershed. They are recognized as home to turtles, snakes, frogs, crocodiles, alligators, waterfowl, mammals, fish, ducks, water lilies and lotus. It provides a home to migrating birds as they feed and rest during their cross-continental passages. But from the 1970s serious loss of species occurred due to overexploitation and deforestation in wetland areas.

3.3 Categories of Wetlands Based on Water

The classification system of wetlands is proposed by Cowardin which he described in the book named "Classification of Wetlands and Deepwater Habitats of the United States". This classification system is used by the Wildlife Service and U.S. Fish programs for the National Wetlands record. This system of classification of the wetlands is classified based on the hydrologic regime and landscape position. It includes 5 types of wetlands namely:

- Lacustrine
- Palustrine
- Riverine
- Estuarine
- Marine

3.3.1 Lacustrine

This is swampy areas within water reservoirs or lakes with low depth and growth of aquatic flora. The word lacustrine means "associated with lakes".

The important characteristics of lacustrine wetland system are:

- Located in a dammed river channel or topographic depression.
- The vegetation lacks shrubs, trees, persistent emergents, lichens, or emergent mosses.
- The total region is of approx. 8 ha.

This type of wetland includes permanently flooded reservoirs and lakes, tidal lakes, and intermittent lakes (Fig. 3.1). Lacustrine type of wetland is divided into 2 sub-systems:

(i) **Littoral zone**:

Fig. 3.1 Lacustrine type of wetland

It is the topmost zone near the shore of lakes. As it absorbs most of the part of sunlight so it is the warmest zone that's why it sustains large communities of fauna and flora. The plant species included in this zone are mostly floating aquatic plants (*Lemna minor*, *Lemna gibba,* etc), rooted plants (*Potamegeton nodusus*, *Nymphea stellata*, *Nymphea alba,* etc) and also the diversity of reptiles, amphibians and fishes occur. Regarding insects, only their eggs and larvae exist in this zone. As the vegetation and animals of this zone provide food to either animals or human beings.

(ii) **Limnetic zone:**

The water surrounds the littoral area is known as the limnetic zone. This is also dominated by zooplankton and phytoplankton same as the littoral zone. In this zone, both the plankton contributed to the food chain.

3.3.2 Palustrine

This type of system contains all nontidal wetlands which are dominated by shrubs, trees, emergent mosses, persistent emergents, lichens, and all such wetlands that occur in tidal regions (Fig. 3.2). It also contains wetlands that lack such type of vegetation, but with all of the following 4 features:

- The total area is less than 8 ha.
- Bedrock shoreline or active wave-formed types lacking.
- The depth of water is not more than 2 m.
- Salinity is less than 0.5%.

This type of wetland was developed to set the vegetated wetlands traditionally called by such names as swamp, marsh, fen, bog, and grasslands. It also consists of the shallow, small intermittent or permanent water bodies commonly called ponds. They are sometimes located at shoreward of river channels, lakes, or estuaries; in isolated catchments; on river floodplains; or hills. In lakes or rivers, they may have occurred as islands. The vegetation nearby lakes and rivers are frequently mentioned as "the shore zone" or the "zone of emergent vegetation". It is also subdivided into the following classes:

- **Aquatic Bed:** This class contains aquatic plants mostly on or under the water surface throughout the growing season this includes pondweeds, floating-leaved plants, water lilies, and bladderworts.
- **Emergent:** Emergent flora exists for most of the growing season. These wetlands contain fens, meadows, and marshes.
- **Persistent:** Those plants that keep standing until the start of the next growing season, e.g. reeds, cattails, bulrushes.
- **Non-persistent:** Species of plant that fall beneath the floor of the water at the end of the developing season so that at some seasons of the year there may be no apparent signal of emergent vegetation, e.g. ferns, arrowheads, and pickerelweed.

3.3.3 *Riverine*

This type of wetland includes all deepwater habitats and wetlands (Fig. 3.3), with two exclusions:

Fig. 3.2 Palustrine type of wetlands

Fig. 3.3 Riverine type of wetland

1. This type is dominated by shrubs, trees, emergent mosses, lichens, persistent emergents.
2. Ocean-derived salts containing water habitats.

This type of wetland is divided into 4 sub-systems. Each subsystem has its characteristic fauna and flora.

- **Tidal:** The gradient is small and water speed varies below the tidal effect. The stream-bed is generally contained mud with random blotches of sand. Deficiency of oxygen may occasionally happen & the fauna is like that in the lower perennial sub-system. It has a well-developed floodplains.
- **Lower Perennial:** Velocity of water is very slow with low intensity of gradient. Water flows throughout the year with no tidal effect. The substrate contains mostly of mud and sand. Oxygen shortages can occasionally arise, the fauna comprised of species that reach their extreme richness in still water, and true planktonic creatures are common. The gradient is much lower compared to the upper perennial subsystem and has well-developed floodplains.
- **Upper Perennial:** Fast speed of water with a high gradient. Very less water flows in the year with no tidal effect. The soil substrate contains cobbles, rock, or gravels with rare spots of sand. Also contains naturally dissolved oxygen concentration. Running water fauna is present, and there are no planktonic forms. Compared to the lower perennial sub-system gradient is high with very little developed floodplains.
- **Intermittent:** Water flows only in one part of the year. When the water is smooth, it remains in isolated pools or the water surface may be absent.

3.3.4 Estuarine

This type consists of deep water tidal territories and neighboring tidal wetlands that are generally semi-enclosed by land but have partly obstructed, open, or sporadic contact to the open ocean, and wherein ocean water is as a minimum sometimes diluted through freshwater overflow from the land. Offshore regions with usual estuarine animals and plants, such as eastern oysters and red mangroves, are also incorporated in this system. This type of wetland is further divided into 2 sub-systems:

- **Subtidal:** In which the soil substrate is always underwater.
- **Intertidal:** In which the substrate is bare and submerged by tides; contains the related splash area.

3.3.5 Marine

The marine system includes the open ocean covering the inland shelf and its related high-power coastline. Marine territories are uncovered to the currents and waves of the water regimes and open-ocean are determined mostly over the flow and ebb of oceanic tides. Salinities go beyond 30%, with no dilution or little excluding the exterior portion of estuaries. Low coastal bays or indentations without considerable freshwater influx and coasts having open stony islands that offer the mainland with no or very little shelter from waves and wind are also considered the portion of the marine system as they usually support characteristic sea biota. In marine system plants and animals are distributed mostly reveals dissimilarities in 4 factors:

- Degree of disclosure of the location to waves.
- Physicochemical nature and texture of the substrate.
- The magnitude of the tides.
- Latitude, which manages the temperature of the water, the duration and intensity of solar radiation, and the absence or presence of ice.

Marine system is divided into 2 subsystems:

- **Subtidal:** In which the soil substrate is always underwater.
- **Intertidal:** In which the substrate is bare and submerged by tides; contains the related splash area.

3.4 Wetlands Types and Classifications

Wetlands have unique features. The water table is the most common characteristic of all wetlands which could be very close to the surface of the soil or shallow water covers the surface for at the least a part of the year. The most important feature of a wetland is decided through the mixture of the salinity of the water inside the wetland, the soil kind, and the flora and animals that exist in the wetland. Based on soil types and mineral contents wetlands are classified into the following types:

3.4.1 Mineral Soil Wetland

This type of wetlands includes Marsh and swamps

- **Marshes**

  This type of wetlands only supports herbaceous vegetation. In this type of wetland, plant life is dominated by sedges and grasses and having poorly drained mineral soil (Lin and Mendelssohn 1996). These commonly exist at the banks of rivers (Fig. 3.4). In marsh places, the flow of water is slower that's why the nutrients enrich sediments are present thus providing circumstances for the further enlargement of the marsh. It is further divided into 2 types:

(i) **Tidal (coastal) marshes**

These types of marshes occur alongside coastlines and are affected by tides and frequently by river water runoff or groundwater. The example of tidal marshes is salt which is considered by salt-tolerant plants. Freshwater tidal marshes are situ-

Fig. 3.4 Vegetation of marshy areas of wetlands

ated upstream of estuaries. The deficiency of salt stress tolerates a larger diversity of vegetation to flourish. Wild rice, cattail, arrowhead, and pickerelweed are commonly grown in these areas and help support various variety of fish species and birds.

(ii) **Non-tidal (inland) marshes**

This type of marsh is dominated by herbaceous plants and commonly occurs in poorly drained floodplains, depressions, shallow water zones, besides the boundaries of rivers and lakes.

- **Swamp**

In this type of wetland plant life is dominated by trees and woody shrubs and having poorly drained mineral soil. Swamps are found in low lying regions connected to the rivers and streams (Fig. 3.5). Some swamps are formed by the marshes that allow trees and the shrubs to grow. They are characterized through very moist soils throughout the developing season & standing water all through some period of the year (Harvey and Hill 2001).

3.4.2 Organic Soil Wetlands

These are also known as "peatlands" which are of two types named Bogs and Fens. These types of wetlands are formed as a result of climatic changes.

- **Bogs:**

This form of wetland having moist spongy surroundings which can be dominated by the vegetation of lower plants like ferns, bryophytes, and mosses. These are regions which can be made because of heavy rainfall (van Zinderen Bakker and

Fig. 3.5 Vegetation of swamp areas of wetlands

Werger 1974). A bog is a wetland in which humidity and dampness are only due to rain or snowfall rather than interior drainage or groundwater (Fig. 3.6).

- **Fens:**

In this category of wetland sedges, reeds and grasses are leading vegetation. These are the regions that are commonly form because of the excessive quantity of water from the surface or groundwater (Van Wirdum 1991). The water ranges in fens stay consistent in the year due to water provided from the underground. In comparison with bogs, fens have less tense circumstances for the development and growth of microbes and plants thus they have well-developed flora.

The dominant genus of plants found in fens wetlands is *Cyperus, Potentilla, Carex*. Fens commonly have a high variety of other plant species including carnivorous plants. Fens also support shrubs, woody plants, and small trees (Fig. 3.7).

3.5 Wetlands of Himalayas

The wetlands that are specifically found in the Himalayan region are the following:

- **Glaciers:** Pakistan has a higher percentage of the glacier-covered area than any other country. The vastest of them is Siachen glacier, having a length of 75 km located in the eastern Karakorum region (Fig. 3.8).
- **Alpine lakes:** There are over 25 important high altitude lakes present in the Himalayan region, usually filled up by glaciers, located at altitudes between 2000 m and over 4000 m. Maximum lakes are situated in the region of Khyber

Fig. 3.6 Vegetation in bog type of wetland

Fig. 3.7 Vegetation in fens type of wetland

Pakhtunkhwa (KPK), Gilgit-Baltistan (GB), and three lakes are present in Azad Jammu & Kashmir (AJ&K). Alpine lakes are categorized by cold, climatic conditions with limited but often unique fauna and flora (Fig. 3.9).

- **Peatlands:** Area covered by peatlands in Pakistan is approximately 20 km^2. Higher altitude peatlands are present in Deosai, Ghizer (Shandur) in Gilgit Baltistan, and Chitral Khyber Pakhtunkhwa (KPK) (Chaudhry 2010).

Fig. 3.8 Glaciers in the Himalayan region

- **<u>Springs and streams</u>**: These are present throughout the country, water collected from the watersheds, and pass on to the rivers with the help of streams and springs. Streams and springs of the Himalayan region contain water from the melting of glaciers (Fig. 3.10).
- **<u>Rivers</u>**: The major river of Pakistan is the Indus which arises from the Himalayas, flows through Gilgit Baltistan, and out from the hills at Attock. Indus river covers the area of 2897 km lengthwise. The rivers flowing from northern areas of Pakistan have cold water as they are formed due to glaciers melting so that's why they have typical vegetation. Rivers are mostly threatened by the unavailability of water due to the construction of Dams and also due to the loss of the flood-plains. The dominant vegetation of these areas is *Hydrilla verticillata*, *Ranunculus*

Fig. 3.9 Alpine lakes in the Himalayan region

trichophyllus, *Potamogeton crispus*, *Shoenoplectus juncoides*, *Acorus calamus*, *Phragmites karka*, *Aconogonon alpinum*, *Equisetum racemosum*, *Carex* spp., *Juncus articulatus*. The accessibility of food resources of wetlands mostly linked to the diversity of vegetation and the production of waste materials (Fig. 3.11).

3.6 Three Important Components of a Wetland

Wetlands have 3 essential components: hydrophytic plants, hydric soils and hydroperiod.

Fig. 3.10 Streams and springs of the Himalayan region

(i) **Hydric soils**

Hydric soils are saturated with water either permanently or seasonally during the growing season and the levels of oxygen become very low which results in anaerobic circumstances in the surface layer of soil. The spaces present in the pores are

Fig. 3.11 Rivers of the Himalayan region

filled with water rather than air. Due to anaerobic circumstances, the plants' growth and their reproduction in the soil become limited. Following are some major characteristics of hydric soils:

- The characteristic "rotten egg" smell of hydrogen sulfide gas linked with anaerobic disintegration.
- Soil color becomes dark especially greys, blacks, or greens.
- Mottling of black or red from mineral concentrations.
- Reddish mottling occurred in rootlets as evading oxygen reacts with iron inside the soil.
- The fine texture of soil reduced the permeability of the soil. Clays and silts are characteristic of wetlands and due to their very fine texture resulting in poor drainage.
- Soils may mire tires, squish underfoot, ooze through fingers, readily form a ball, or show standing water in holes.

(ii) **Hydrophytic plants**

The term "Hydrophytic" exactly means water-loving. These types of plants employ several adaptations for existence in an anaerobic and waterlogged environment. Terrestrial plants naturally uptake oxygen by roots. Wetland flora is not able to do that so that they have evolved captivating alternatives to acquire and transport oxygen. Adaptations of these plants include:

- Roots and stems having hollow tubes called aerenchyma which helps to transport oxygen as well as provides physical support to these plants. These hollow tubes are found in sedges, grasses, and rushes.

- Plants that grow in shallow water or emergent species with tree's root system are known as "knees." for example, these knees help a bald cypress tree to get oxygen from the topmost aerobic layers of soil or atmosphere. Cypress knees frequently appear like thick foot-excessive woody palms extending above the waterline across the base of a tree.
- Wetland vegetation is light and floats on the surface of the water. A common example of which is Duckweed. With leaves which can be much less than ¼ inch throughout and roots that hang withinside the water, duckweeds also are the smallest higher plants (has roots, stems, and leaves) in the world.
- Trees in swamplands frequently show "buttressed" trunks. Buttressed trunks are flared out or thickened nearby the base of the tree. This aids largely to support the tree in saturated soils that offer slight support. High levels of water are usually indicated by the height of these buttressing.
- Many trees that grow in wetlands their roots are often visible on the surface of the soil. These characteristic roots allow oxygen to absorb and also they provide stabilization to the trunk. The most common example is the willow tree.

(iii) **Wetland hydroperiod**

Hydroperiod refers back to the sample of water ranges in anyone specific wetland. The level of water in wetlands may differ seasonally, daily, or more unevenly. Alternately, wetlands can be completely wet. In any case, the hydroperiod is an essential character of every wetland. It is the occurrence of water near or at the surface of the soil which causes soil anaerobic conditions and saturation that benefits the hydrophytic vegetation.

3.7 Classification of Wetland Flora

Wetland vascular plant species are typically categorized primarily based totally on their growth form (Boutin and Keddy 1993). It is based solely on how the plants grow in physical relationship to the water and soil. We follow in adopting the simplest scheme with the least amount of terminology. The categories used to group wetland plants include the emergent, submerged, floating, and moist places plants. The general characteristics of each group are described below.

(a) **Emergent**

Emergent plants are rooted in the soil with basal portions that typically grow beneath the surface of the water, but whose leaves, stems (photosynthetic parts), and reproductive organs are aerial (Fig. 3.12). The most common emergent species are found in the large families of monocotyledons that tend to dominate both freshwater and saltwater marshes, i.e., the Poaceae (grasses), Cyperaceae (sedges, e.g., *Carex*, *Cyperus*), Juncaceae (rushes), and the Typhaceae (cattail). Other families with frequently encountered emergent species are the Alismataceae (water plantain), Araceae (arum), Asteraceae (aster), Lamiaceae (mint, e.g., *Lycopus*, *Mentha*), Polygonaceae (smartweed), and Sparganiaceae (bur reed) (Vymazal 2013).

Fig. 3.12 Pictorial view of Emergents

Fig. 3.13 Pictorial view of Submerged plants

(b) **Submerged Plants**

With the possible exception of flowering, submerged plants typically spend their entire life cycle beneath the surface of the water and are distributed in coastal, estuarine, and freshwater habitats (Fig. 3.13). Stems and leaves of submerged species tend to be soft (lacking lignin) with leaves that are either elongated and ribbon-like or highly divided, making them flexible enough to withstand water movement without damage (Dogan et al. 2009). Examples of families in which all or nearly all of the species are submerged include the Callitrichaceae (water starwort), Ceratophyllaceae (hornwort), Haloragaceae (watermilfoil), Potamogetonaceae (pondweeds), and Lentibulariaceae (bladderworts).

(c) **Floating Plants**

The leaves and stems of floating plants (also known as floating unattached) float on the water's surface (Fig. 3.14). If roots are present, they hang free in the water and are not anchored in the sediments (Meerhoff et al. 2003). A widespread family of free-floating plants is the Lemnaceae, which includes the genera *Lemna* (duckweed), *Spirodela* (greater duckweed), and *Wolffiella* and *Wolffia* (watermeal).

(d) **Amphibious plants**

This category includes those plants which need wet areas to grow including bryophytes pteridophytes and some angiosperms also (Fig. 3.15). These types of plants grow in moist places where neither abundant water nor dry situations. They are found along the margins of water-bodies. These plants can either tolerate or enjoy moist to wet garden conditions. Some perennials will adapt to these conditions provided that the soil dries while dormant in winter, but these plants tend to thrive in moist conditions. These plants are either rooted in mud or grow up in the water or out in the air (Bernhardt et al. 2008).

Fig. 3.14 Pictorial view of Floaters

Fig. 3.15 Pictorial view of Amphibious plants

References

Bernhardt KG, Koch M, Kropf M, Ulbel E, Webhofer J (2008) Comparison of two methods characterising the seed bank of amphibious plants in submerged sediments. Aquat Bot 88(2):171–177

Boutin C, Keddy PA (1993) A functional classification of wetland plants. J Veg Sci 4(5):591–600

Chaudhry AA (2010) Wetlands in Pakistan: what is happening to them. World Environment Day, 5

Dogan OK, Akyurek Z, Beklioglu M (2009) Identification and mapping of submerged plants in a shallow lake using quickbird satellite data. J Environ Manag 90(7):2138–2143

Harvey KR, Hill GJ (2001) Vegetation mapping of a tropical freshwater swamp in the Northern Territory, Australia: a comparison of aerial photography, Landsat TM and SPOT satellite imagery. Int J Remote Sens 22(15):2911–2925

Lin Q, Mendelssohn IA (1996) A comparative investigation of the effects of South Louisiana crude oil on the vegetation of fresh, brackish and salt marshes. Mar Pollut Bull 32(2):202–209

Meerhoff M, Mazzeo N, Moss B, Rodríguez-Gallego L (2003) The structuring role of free-floating versus submerged plants in a subtropical shallow lake. Aquat Ecol 37(4):377–391

Van Wirdum G (1991) Vegetation and hydrology of floating rich-fens. Datawyse, Maastricht

van Zinderen Bakker EM, Werger MJ (1974) Environment, vegetation and phytogeography of the high-altitude bogs of Lesotho. Vegetatio 29:37–49

Vymazal J (2013) Emergent plants used in free water surface constructed wetlands: a review. Ecol Eng 61:582–592

Chapter 4
Importance of Biodiversity in Wetlands

4.1 Importance of Biodiversity

The variety of life on earth is generally termed as biodiversity. It includes different number of animals, plants and microbes species. It also comprised of organisms from different ecosystems, containing rainforests, deserts, grasslands, coral reefs, polar ice caps and tundra. This biodiversity plays vital role for the well-being of our world. Most of cultures in the world are familiar with the importance of protecting natural assets. Several still do, but several do not.

4.2 Benefits of Biodiversity

Function of ecosystem and services provide from it totally depends on the biodiversity. List of benefits and services provide by biodiversity is mentioned below:

- Provisioning facilities such as clean water, food, timber, genetic resources and fiber.
- Regulating services such as floods, climate, water quality, disease and pollination.
- Traditional facilities such as aesthetic, recreational and spiritual welfares
- Supporting facilities such as nutrient cycling and soil formation.

M. A. Butt et al., *Wetland Plants*, https://doi.org/10.1007/978-3-030-69258-2_4

4.3 Types of Biodiversity

There are three main types of biodiversity:

- Genetic biodiversity
- Species biodiversity
- Ecosystem diversity

4.3.1 Genetic Biodiversity

The alternate concept of biodiversity is generally termed as Genetic diversity. Genetic variation is the raw material for adaptation in a species. A species' future potential for adaptation depends on the genetic diversity held in the genomes of the individuals in populations that make up the species. The similar is accurate for higher groups of plants. A genus with variation in species will have more genetic diversity than a genus with similar species and have similar ecologies. The genus having maximum potential for successive evolution is genetically diverse one.

At present, it is far cheaper to determine chemical compounds which are made by organisms than to manufacture them in a controlled condition. The diversity of chemical is one way to measure diversity that is significant to human welfare and their health. Through selective breeding, people have domesticated plants, fungi and animals, however even this range is struggling losses due to marketplace forces and growing globalism in human migration and agriculture. For an instance, global seed corporations produce only a very few sorts of a given crop and offer incentives round the globe for farmers to shop for those few types whilst leaving behind their conventional types, which might be a ways greater diverse. The human populace relies upon on crop range at once as a solid meals supply and its decline is disturbing to agricultural and biological scientists.

4.3.2 Ecosystems Diversity

Ecosystem diversity is defined as "the number of different ecosystems on Earth or in a geographical area". Whole ecosystems can disappear even though a number of the species may continue to exist through adapting to different ecosystems. Due to lack of interaction between the species the whole ecosystem cause damage also involves the loss of distinctive features of co-adaptation, and biological productivity loss that an ecosystem is able to generate. According to an example in North America a largely extinct prairie ecosystem is due to these factors, which covered central North America from the boreal forest in northern Canada down into Mexico. Which are disappeared and converted into crop fields, suburban sprawl and pastures. Many of the species survive, however the highly efficient atmosphere that became accountable for developing our maximum efficient agricultural soils is now gone. As a consequence, their soils are actually being depleted until they may be maintained artificially at more expense. he decline in soil productiveness takes place due to the fact the interactions in unique surroundings were lost; this became a much greater critical loss than the quite few species that had been driven to extinct when the grassland ecosystem was demolished.

4.3.3 Species Diversity

Diversity of species is defined as the variety of species within a habitat or a region. They are the basic units of taxonomy and thus the common amount of biological diversity. Another term species richness is used which describes the number of different species in a given area. The world total is estimated at 5–10 million species, though only 1.75 million have been named scientifically so far.

4.4 Importance of Biodiversity in Wetlands

Wetland biodiversity can be defined as the variety of life and the ecosystems that make up the freshwater, tidal, and marine regions of the world and their interactions. Wetland biodiversity consists of two types of ecosystems

- Freshwater ecosystems (ponds, lakes, rivers, reservoirs, groundwater, streams and wetlands).
- Marine ecosystems (estuaries, oceans, seagrass beds, salt marshes, kelp beds, coral reefs and mangroves.

Wetland biodiversity includes all unique species, their habitats and interaction between them. It consists of zooplankton, phytoplankton, insects, aquatic plants, birds, fish, vertebrates, invertebrates and others. Biodiversity in wetland has huge aesthetic and economic values and is mostly responsible for supporting and maintaining overall health of environment. People have depended on aquatic resources for medicines, food, shelter, materials as well as for commercial and recreational purposes such as tourism and fishing. The wetland fauna and flora depends upon the great variety of wetland habitats and means for food, materials, and breeding grounds.

The factors which are continuously declining wetland biodiversity including both marine and freshwater is overexploitation of species, exotic species introduction, water pollution, agricultural and industrial wastes, habitat loss and conversion of wetland habitats through damming and water diversion. As a result, precious aquatic assets have become an increasing number of at risk of each artificial and natural environmental change. Thus, conservation approaches to conserve and protect aquatic life are required to keep the stability of nature and support the availability of resources for future generations.

4.5 Importance of Wetland Flora

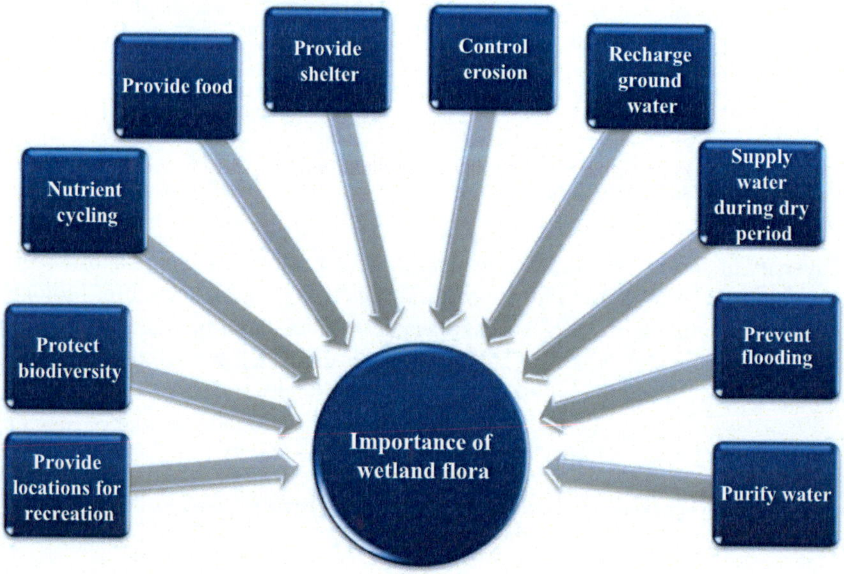

4.5.1 Purify Water

Wetlands act as great filters of nature; they capture the sediments and remove water pollutants, which aid in water purification. This actually beats costly, man-made systems of water filtration. Wetlands helps infiltration of water by eliminating extra nutrients, slow down the flow of water due to which particles settle down in the bottom and absorbed by the roots of plants. It is proven in previous research that up to 95% of nitrogen and 92% of phosphorus can be eliminated from water while passing through the wetland. They allow the particles to stick to soil and release pure water. Research has been done on some wetland plants which have capacity to hold heavy metals from the water and let the water purify. Without those functions, the waterways might constantly increase their nutrient and pollutant load, which results in remoted deposit of excessive concentrations in addition down the line. Such an example is found in Mississippi River's dead zone, where accumulation of excess nutrients occur that has led to the production of large amounts of surface algae, which utilize oxygen and form hypoxic conditions.

Another benefit of wetlands is they absorb and filter out harmful bacteria from the water. Their complex food chain hosts various bacteria and microbes, to which invertebrates used as food and in return they filter up to 90% of bacteria out of the water.

4.5.2 Store Water

Wetlands function like large sponges. They keep water after which gradually release it and this facilitate to cope with dry seasons with slight rainfall.

4.5.3 Prevent Floods

Due to heavy rainfall rivers burst their banks, in this condition wetlands can save the additional water, and slow down the water so it distributes uniformly over a floodplain. Also the vegetation across the river also helps to slow down the speed of flood waters.

Due to thick flora and site within the land, wetlands are significant for holding storm water from melting snow and rain water moving closer to floodwater and surface water from over flowing watercourses. Wetlands slow down the motion of water from the storm and can provide storage regions for floods, thus reducing negative effects to downstream regions. Conservation of wetlands can prevent unnecessary costs for storm and floodwater control plans such as levees, dikes, detention basins, and concrete-lined channels.

Wetlands which are located in the lower or mid reaches of a watershed make contributions maximum considerably to flood manage considering they're with inside the route of greater water than their upstream counterparts. When numerous wetland basins carry out this feature inside a watershed, the impact can be a moderated discharge, staggered, decreasing flood peaks.

Flood safety can be mainly critical in urban settings and regions with steep slopes, overgrazing, or different land functions which have a tendency to raise storm water quantities and velocity. These useful values can provide financial aids to downstream tax payers and property owners.

4.5.4 Recharge Ground Water

In earlier days, wetlands are filled or in the place of these wetland dams have been constructed, addition of pipes that leads water to the ocean as fast as possible. At present it is known that wetlands allow water to immerse into the ground, and to refill the natural ground-water supply.

4.5.5 Control Soil Erosion

In wetlands the pollutants are trapped. In semi-arid regions of the world, sediments are usually trapped in wetlands, after this these sediments joins a river channels and washes away, which is actually beneficial. Emergent and submerged plants defend the water surroundings from soil erosion due to wave currents or action. These plants also stabilize the pollutants which can increase clarity of water.

4.5.6 Provides Shelter

Shallow water of wetlands is used as nursery for some fish like Fish larvae and juveniles. Some amphibians and young fish use wetland vegetation as a source of cover from predatory birds and fish. These plants also provide food for baby frogs, salamanders and fish. In the surrounding of water there is high amount of biodiversity. Vegetation on the surrounding of water bodies not only provide shelter but also protect wetland animals and reptiles for predators, they also lays eggs in these plants and some plants are rich in nutrients so animals also feed them.

Wetland flora provides living space for minor animals such as insects, crustaceans and snails which in turn supply food for fish (Fig. 4.1). In previous studies it is reported that the area with dense vegetation has more fauna as compared to less vegetative areas. Commonly used plants that provide food and shelter to insect, amphibians, birds, fish and insects are *M. spicatum* and *C. dermesum*. The root of floaters such as *S. nymphellula* and *Pistia stratiotes* provides shelter and hiding place for invertebrates and fish from predation by other animals. Some fish that build their nests to lay eggs they use submerged plants to make nests whereas Elephant trunk fish only breeds in still waters which is covered with submerged plants but after breeding they moves to more flowing body of water.

4.5.7 Wetlands Provide Food for Humans and Livestock

Wetlands support grasses to grow which provide good areas for grazing, and can be useful to farming livestock. Wetland vegetation provides essential food for many animals. Geese and ducks eat the leaves, tubers and seeds of many wetland plants such as *Brasenia schreberi*, *Potomogeton* spp., *Lemna* spp., *Polygonum* spp. and *Sagittaria latifolia*. Some birds use *Typha* fur to make their nests and also eat their seeds. Beaver, Otter, turtles, moose and muskrats, also use a variety of submerged and floating plants as a source of their food (Fig. 4.2).

In earlier days many people throughout the world use water plants as a source of food. *Typha* spp. have eatable roots and shoots, even the pollen has been used in making soups and biscuits. *Sagittaria* tubers are known as duck potatoes, which

Fig. 4.1 Diversity of fauna in wetlands

were consumed by many people. *Nasturtium officinale* is used as medicines in earlier time also it continues to be used in salads and cooked as vegetable. Roots of *Nymphea* are used as medicine as well as it is used as food in many parts of the world. *Ceratophylum demersum* a submerged plant has been used for medicinal purposes.

We use a wealth of herbal products from wetland, which includes shellfish and fish, cranberries, blueberries, wild rice and wood. Some herbal medicines also

Fig. 4.2 Grazing of wetland flora by life stock

derived from wetland plants and soils. Many of the countries shell fishing and fishing industries yield wetland-established species. In the Southeast, for example, almost all of the business catch and over half of the leisure harvest are shellfish and fish that depend upon the estuary-coastal wetland system.

4.5.8 Protect Biodiversity

Many unique types of creatures depend upon wetlands– and upon one another. Vegetation of wetlands attract variety of insects which are the food source for animals like frogs, birds and fish which in return attract other predators. In wetlands the biodiversity has capacity to produce some unique individual species which can be best located in those habitats.

4.5.9 Locations for Recreation

Ratti Gali lake, Chita Katha lake, Mahundand lake are the examples of wetlands of Himalayas where millions of tourists visit and enjoy nature walks, birding, picnics, fishing or even boating. As cities are crowded and noisy so many people came to these recreational spaces to relax and spend holidays. Also the diverse flora of wetland plants including emergents, submerged and floaters, particularly those plants having brightly coloured flowers, adds overall beauty of these sites. The most important benefit of growing healthy network of local wetland plants is to reduce the value of exotic invasive plant species to colonize pond or lake. Balance in wetland plants life results in better and healthy aquatic fauna, such as plankton and fish.

Another aspect which should be keep in mind while maintaining healthy wetland native flora that it needs energetic control so that to get the maximum recreational use out of the waterbody. Due to excessive growth of emergents they can cause damage to lower plants, and due to completely covered floating plants it can cause unsatisfying fishing conditions. Wetlands offer excellent academic and clinical studies possibilities due to their specific combination of aquatic and terrestrial existence and chemical/physical processes. Many species of threatened and endangered flora and fauna are discovered in wetlands. Wetlands positioned inside or close to city settings and those often visited by the people are of great importance as they are offering educational and social opportunities. Open water, numerous vegetation, and absence of pollutants additionally make a contribution to the price of unique wetlands for leisure and academic functions and popular exceptional of existence. Also in earlier days wetlands are used by poets, musicians, book writers for creative thoughts as these areas provide relax piece of mind for thoughts.

4.5.10 Houses and Crafts Making

Plants that grow in wetlands are said to be good provider of fuelwood for food preparation, roofing, and fibres for paper making and used in textiles, and timber for construction purposes. Many parts of wetland plants such as (leaves, fruits and barks) are used as medicinal purposes, they also used in dyes, staining and tannins, also in treatment of leather and leather products. The strong emergents make available for bird nesting and in making shelters for many insects, birds, reptiles, mammals and amphibians. From emergent plants brooms, baskets, pillows, mats, boats, and even houses are made from the stems of rush, cattail and bulrush.

Since ancient time's people use some species of *Cyperus* genus to weave hats, mats, spoon holders, bags and other instruments for their own usage. The humans gather the needful weaving substances immediately from the limnetic water bodies and from marshes. After survey of these areas some researchers reported that there is extreme decrease in the population of *Cyperus* species because of their overexploitation in the weaving industry. A part from this some researchers also reported that some species of *Typha* is also used for weaving materials.

4.5.11 Biological Productivity

Due to highest capacity of water absorption wetlands are highly biologically productive (capable of producing biomass rapidly). On the basis of plant productivity freshwater wetlands are like to tropical rainforests. Their capacity to effectively create biomass may become significant to the progress of alternate energy sources. Fresh water contains excess amount of algae which helps in the production of biomass.

4.5.12 Nutrient Cycling

Wetland vegetation plays an important part in complex system of chemical cycling which occurs in waterbody. These plants also affect the oxygen supply in the water. In recent times these plants have gained a lot of consideration due to their capacity to absorb pollutants from polluted water. They consume nutrients that might in any other case be utilized by algae, thereby enhancing the clarity of water. Increase in consideration helps to use them as indicators of water quality.

Wetland plants are the keystones of any wetland ecosystem and they have physiological capability of eliminating heavy metals and mineral nutrients (Uka et al. 2009). The study of wetland plants is a necessary factor of understanding a water body because of its vital role in ecology and its capacity to describe the quality of water (Ghavzan et al. 2006). Wetland plants are an essential element in complex system of chemical cycle and they affect the amount of oxygen supply in the water (Table 4.1).

Table 4.1 Removal efficiency of metal ions by some common aquatic macrophytes

Aquatic macrophytes	Metals	Removal efficiency
Azolla pinnata	Hg	93.0
Ceratophyllum demerum	Pb, Zn, Cu	80.0
Echhornia crassipes	Fe, Cu, Zn. Cd	80.0
Ipomea aquatica	Hg	90.0
Lemna minorl	Pb	90.0
Ludwigia repens	Hg	99.0
Pistia stratiotes	Cd, Hg, Cr	85.9

Source: Srivastava et al. (2008)

Many researchers throughout the world have been worked on the chemical cycle of wetland plants. Boyd (1970a) revealed that vascular wetland plants are utilized for the deletion of pollutants from polluted or eutrophic water bodies specifically where these pollutants are near the fish ponds. According to Agbogidi et al. (2000), the waste materials are passed over the field having populations of wetland plants which reduce and absorb the pollutants before going into the pond. Boyd (1970b) suggested that this has been utilized in bio-manipulation of fish ponds to improve production of fish. Dar-Nimrod and Heine (2011) observed the phyto-remediative ability of water hyacinth for treating wastewater from homes. The outcomes confirmed that the experimental plant decrease all biological and physicochemical parameters to a huge level. Thus it may be used as an economic, efficient and ecological substitute to speed up the degradation and removal of waste water pollutants from agro-industry. Chantiratikul et al. (2008) performed research at the feasibility of manufacturing selenium-enriched water lettuce for animal nutrients in hydroponics. The end result confirmed that selenium concentrations in roots and leaves of water lettuce expanded substantially with growing concentration of Se in the solution and exposure time, therefore water lettuce have excessive capability in steel absorption. Another research on passive phyto-remediation was conducted at Ologe Lagoon, Lagos-Nigeria. The research assumed that *Eichhornia crassipes* can gather heavy metals even if the metals concentrations in sediment and water of the aquatic environments is low signifying that *Eichhornia crassipes* may be used to remediate polluted water (Ndimele and Jimoh 2011). Another study conducted by Boyd (1970a) reported the loss of nutrients from the leaves of *Typha latifolia* which is an emergent plant. Wetland plants absorb huge quantity of nutrients and which is an effective way of eradicating nutrients from natural waters or effluents (Uka et al. 2009). Brix and Schierup (1989) stated in their study that internationally the attention of people is now towards the capability of macrophytes to control the water pollution and also to treat industrial and municipal wastewater. Thus, wetland plants can be able to indicate the quality of water and their existence may improve the quality of water due to their capacity to absorb unnecessary burden of nutrients (Petre 1990). It has been revealed that wetland weeds such as *Lemna* sp., *Eichhornia crassipes*, *Typha* ssp. *Salvinia* spp., *Azolla* sp. and *Phragmites* spp. has been used as a way of dropping the levels of nutrient of polluted water (Oki et al. 1989), Whereas *Phragmites karka* has little heavy metal isolating capacity (Uka and Chukwuka 2011).

4.5.13 Floral Diversity

Wetlands can support wealth and diversity of flora, starting from orchids and duckweed to black ash. This flora makes contributions to the earth's biodiversity and offer food and provides home for lots faunal species at important times throughout their living cycles (Fig. 4.3).

In many countries many endangered plant species are found in wetlands. The significance of plants diversity in a specific wetland is generally associated with two factors.

- Firstly, the valued wetlands generally support a variety of local flora, instead of sites with slight low diversity of non-native flora.
- Secondly, wetlands accommodating those communities of plants which are locally uncommon and are deliberately valuable.

Nelumbo nucifera flowers are used in many countries for decoration and used during prayers in temples. *Nymphaea pubescens* flowers are also used in temple offerings. Though, the existence of *Nymphaea nouchali* is rare, and as a result is less collected for offerings. The requirement for plants will increase at some stage in non-secular seasons like Poson and at some stage in wedding ceremony seasons due to the high demand for flower arrangements flower offerings respectively, during such times (Table 4.2).

Fig. 4.3 Floral Diversity of wetland flora

Table 4.2 Economically important some wetland plants

Sr. no.	Plant name	Common name	Family	Part used	Economic importance	References
1.	*Alternanthera philoxeroides* (Mart.) Griseb.	Alligator weed	Amaranthaceae	Whole plant	This plant is used as animal feed. It contains about 8.2–15.0% dry matter and 8.1–12.8% protein on dry weight basis.	Rahman and Gulshana (2014), Dutta (2015)
2.	*Aponogeton natans* (L.) Engler et. Krause.	Floating lace plant	Aponogetonaceae	Leaves, seeds and young shoots	Decoction of leaves is used in stomach disorder and digestive problems. Starchy seeds is roasted and taken as food. Flowering spike and young shoots are used as vegetables.	Dash et al. (2017)
3.	*Asteracantha longifolia* (L.) Nees.	Asteracantha	Acanthaceae	Whole plant, seed	Entire plant is used in anaemia, jaundice, rheumatism, to stop bleeding and to improve sexual power. Seed pest is used in tubercular fistula. Decoction of leaves used to treat dropsy. Root pest is applied externally in rheumatism.	Chauhan and Dixit (2010)
4.	*Azolla pinnata* R. Br.	Mosquito fern	Salviniaceae	Whole plant	This delicate, small aquatic plant is of considerable economic importance and extensive researches have been conducted with it. It has been reported useful for controlling mosquitoes and other weeds, and also for feeding poultry, pigs and ducks. However, Azolla's greatest economic value lies in its nitrogen fixing capacity, whereby the plant is capable of reducing the nitrogen demand in paddy fields. Azolla is used as a biofertiliser for increasing the yield of rice in China, Vietnam, Philippines, India and some other countries.	Mithraja et al. (2011)
5.	*Bacopa monnieri* (L.) Wettst.	Water hyssop	Plantaginaceae	Whole plant	This plant is also a good source of human food and the juice is used to cure fever.	Kean et al. (2017)

6.	*Cynodon dactylon* (L.) Pers.	Bermuda grass	Poaceae	Whole plant	It is a good source of food to the aquatic molluscans. This plant is used to cure skin diseases. People utilize this plant to worship god.	Shendye and Gurav (2014)
7.	*Cyperus exaltatus* Retz.	Giant sedge	Cyperaceae	Whole plant	Poultry feed, perfume and insecticides. Used for paper making.	Sivapalan (2013)
8.	*Cyperus iria* L.	Flat sedge	Cyperaceae	Stem, roots	The stem of the plant has been used for writing material, paper, since ancient times. The plant is also a source of fiber, fuel and food. The roots are used as fuel and the pith made into food.	Ho et al. (2012)
9.	*Dentella repens* (L.) J.R.Forst. & G.Forst.	Creeping lickstoop	Rubiaceae	Leaves	Leaves are used as blood purifier, to improve the eyesight and for poultice sores.	Das et al. (2016), Panda (2006)
10.	*Echinochloa colona* (L.) link	Jungle rice	Poaceae	Whole plant	Used as fodder for life stock	
11.	*Hydrilla verticillata* (L.f.) Royle	Water thyme	Hydrocharitaceae	Whole plant	Hydrilla has shown promise as a source of biogas. This plant is additionally rich in carotene and xanthophylls. It is also used as a fish feed. Moreover, the plant is used by fish farmers to breed since the eggs are adhered to this plant.	Pieterse (1981)
12.	*Ipomoea aquatica* Forssk.	Water spinach	Convolvulaceae	Leaves, stem, whole plant	The young leaves and stems are used as vegetable, boiled or cooked in oil. It is also used for pickles preparation and also as animal feed.	Malakar and Choudhury (2015)
13.	*Ludwigia adscendens* (L.) H. Hara	Water primrose	Onagraceae	Whole plant	Dried plant powder is applied externally on skin to cure various skin diseases. Also used as anti-fertility drugs.	Chakraborty et al. (2014)

(continued)

Table 4.2 (Continued)

Sr. no.	Plant name	Common name	Family	Part used	Economic importance	References
14.	*Monochoria vaginalis* (Burm.f.) C.Presl	Water-hyacinth	Pontederiaceae	Whole plant	It is good source of organic fertilizer. Liquid extracts from the plant improved the seed viability rate and increase the seedling vigour of some crops. It is also a source of cattle feed.	Chandran et al. (2012)
15.	*Nelumbo nucifera* Gaertn.	Lotus	Nelumbonaceae	Seeds, flowers, rhizome	The seeds and the rhizomes are used in a variety of cooked and uncooked dishes. The seeds were also ground to flour for making bread. Rhizomes are marketed fresh, dried, canned or even as a fine powder. They are cooked in curries and other oriental dishes. The dried and canned rhizomes, both sell at a high price. The seeds are eaten raw, cooked, ground to flour or canned after removing the skin and the bitter embryo. The petals are considered as cardiac tonic and said to have a cooling effect. The lotus is also used as an ornamental plant in this area.	La-Ongsri et al. (2009), Tungmunnithum et al. (2018)
16.	*Nymphaea nouchali* Burm.f.	Blue lotus	Nymphaeaceae	Petals, rhizome	The spongy petioles and peduncles are cooked and eaten by the poor people. The powdered rhizome is used in dyspensia, dysentery and piles. The edible seeds may be eaten raw or roasted in sand before eating.	Raja et al. (2010)
17.	*Nymphoides indica* (L.) Kuntz.	Water snowflake	Menyanthaceae	Leaves	Leaf pest is mixed with water and drunk once a day to cure jaundice, fever and dysentery. Leaf pest is applied on forehead to get relief from headache. Leaf pest is externally applied to the swelling part of the body to get relief from pain.	Madhavan et al. (2009)
18.	*Oryza sativa* L.	Asian rice	Poaceae	Whole plant	It is the marshy aquatic plant, which is used as human food crop. It is a stable food in areas often submerged by flood water. This plant is used as an animal fodder to cows and buffaloes.	Cabanting and Perez (2016)

19.	*Pistia stratiotes* L.	Water lettuce	Araceae	Whole plant	Whole plant is used as feed for fish, ducks and reptiles.	Khan and Arshad (2014)
20.	*Ranunculus sceleratus* L.	Celery-leaved buttercup	Ranunculaceae	Rhizome, seeds, leaves, shoots	One tea spoon of rhizome powder is taken with water to cure diarrhea and dysentery. One table spoon juice of leaf extract is mixed with a cup of water and takes once a daily for a week to remove ringworm. Seeds are used as tonic and in kidney troubles. The young shoot and roots are cooked with vegetables.	Mei et al. (2012)
21.	*Salvinia natans* (L.) all.	Water moss	Salviniaceae	Whole plant	Used for life stock fodder.	Al-Maliki et al. (2017)
22.	*Spirodela polyrrhiza* (L.) Schleid.	Duck meat	Araceae	Stem, leaves, whole plant	Plant has no distinct stem or leaves, but consists of leaf like fronds. These plants are collected in huge quantities and used as organic manures or fodder for cattle and pigs in tropical regions of India. The plants are also consumed by herbivorous fish, ducks, goose and swans, etc. the high nitrogen value of the duckweed (6–7%) makes the plant an ideal source of organic manures. High protein content makes it as a source of animal feed. Moreover, this plant may also be used for purifying wastewater. The liquid extract from Spirodela was found to improve seed viability and increase seedling vigour of maximum number of seeds.	Bolotova (2015)
23.	*Trapa natans* var. *bispinosa* (Roxb.) Makino	Water chestnut	Lythraceae	Fruit	Fruit a spiny nut, with 4 prominent angles is edible. The fruits are harvested from the naturally growing plants and sold in market. The fruits are eaten raw or sweet dishes are prepared with them.	Abubakr et al. (2011)

(continued)

Table 4.2 (Continued)

Sr. no.	Plant name	Common name	Family	Part used	Economic importance	References
24.	*Typha* species	Cattail	Typhaceae	Fruit, pollen	The cattails are fast growing and invade waterways, rice fields, farm ponds, lakes, tanks and canals. Of all the wild plants, cattails have been called the most useful emergency food source. Traditionally these are important food to native people. They contain much protein like corn or rice and more carbohydrate than potato. In times of food scarcity, cattails have been considered as a source of starch, the seeds a source of edible oil and animal feed. Young cattail stems may be eaten as salads or as green vegetables; even the yellow pollen may be eaten. Typha species is a promising source of pulp and paper. The plants absorb nutrients from polluted wastewaters, purifying the water during the process.	Bansal et al. (2019), Liping and Hui (2007)
25.	*Vallisneria spiralis* L.	Eel grass	Hydrocharitaceae	Whole plant	It is a submerged plant, used for ornamental purpose. This plant is also used by fish culturists to breed the fishes. The cyprinid fish and grass carp are rapid feeder of eel grass.	Rai and Tripathi (2009)

References

Abubakr A, Nahvi A, Kundangar MRD (2011) Ecological studies and uses of valued aquatic plants in Kashmir wetlands. Nat Environ Pollut Technol 10(1):115–118

Agbogidi OM, Bamidele JE, Ekokotu PA, Olele NF (2000) The role and management of aquatic macrophytes in fisheries and aquaculture. Issues Anim Sci 10:221–235

Al-Maliki GM, Al-Khafaji KK, Karim RM (2017) Antibacterial activity of two water plants Nymphaea alba and Salvinia natans leaves against pathogenic bacteria. Int J Fish Aquat Stud 5(5):353–355

Bansal S, Lishawa SC, Newman S, Tangen BA, Wilcox D, Albert D, Anteau MJ, Chimney MJ, Cressey RL, DeKeyser E, Elgersma KJ (2019) Typha (Cattail) invasion in north American wetlands: biology, regional problems, impacts, ecosystem services, and management. Wetlands 39(4):645–684

Bolotova YV (2015) Aquatic plants of the Far East of Russia: a review on their use in medicine, pharmacological activity. Bangladesh J Med Sci 14(1):9–13

Boyd CE (1970a) Chemical analyses of some vascular aquatic plants. Arch Hydrobiol 67:78–85

Boyd CE (1970b) Vascular aquatic plants for mineral nutrient removal from polluted waters. Econ Bot 24(1):95–103

Brix H, Schierup HH (1989) The use of aquatic macrophytes in water-pollution control. Ambio 28(2):100–107

Cabanting RMF, Perez LM (2016) An ethnobotanical study of traditional rice landraces (Oryza sativa L.) used for medical treatment in selected local communities of the Philippines. J Ethnopharmacol 194:767–773

Chakraborty I, Ghosal S, Pradhan NK (2014) Jussiaea repens (L) acts as an antifertility agent–a search for herbal male contraceptive. Int J Pharm Sci Rev Res 24(2):288–296

Chandran R, Thangaraj P, Shanmugam S, Thankarajan S, Karuppusamy A (2012) Antioxidant and anti-inflammatory potential of Monochoria vaginalis (burm. f.) c. presl.: a wild edible plant. J Food Biochem 36(4):421–431

Chantiratikul A, Chinrasri O, Chantiratikul P (2008) Effect of sodium selenite and zinc-L-selenomethionine on performance and selenium concentrations in eggs of laying hens. Asian Australas J Anim Sci 21(7):1048–1052

Chauhan NS, Dixit VK (2010) Asteracantha longifolia (L.) Nees, Acanthaceae: chemistry, traditional, medicinal uses and its pharmacological activities-a review. Rev Bras 20(5):812–817

Dar-Nimrod I, Heine SJ (2011) Genetic essentialism: on the deceptive determinism of DNA. Psychol Bull 137(5):800

Das D, Mondal S, Mandal S (2016) Studies on some economically important aquatic plants of Katwa subdivision of Burdwan District, West Bengal, India. Int J Curr Microbiol App Sci 5(6):961–972

Dash S, Sahoo AC, Senapati AK, Sahoo PK (2017) Acute toxicity study of Aponogeton natans-an important folklore medicine. J Chem Pharm Sci 10(1):626–629

Dutta P (2015) Pharmacognostical evaluation and preliminary phytochemical analysis of Alternanthera philoxeroides. Int J MediPharm Res 1(1):7–13

Ghavzan NJ, Gunale VR, Mahajan DM, Shirke DR (2006) Effects of environmental factors on ecology and distribution of aquatic macrophytes. Asian J Plant Sci 5(5):871–880

Ho YL, Huang SS, Deng JS, Lin YH, Chang YS, Huang GJ (2012) In vitro antioxidant properties and total phenolic contents of wetland medicinal plants in Taiwan. Bot Stud 53(1):55

Kean JD, Downey LA, Stough C (2017) Systematic overview of Bacopa monnieri (L.) Wettst. dominant poly-herbal formulas in children and adolescents. Medicines 4(4):86

Khan AA, Arshad SANA (2014) Wetlands of Pakistan: distribution, degradation and management. Pak Geogr Rev 69(1):28–45

La-Ongsri W, Trisonthi C, Balslev H (2009) Management and use of Nelumbo nucifera Gaertn. in Thai wetlands. Wetl Ecol Manag 17(4):279–289

Liping Z, Hui X (2007) Comprehensive development and uses of the wild plant resource Typha [J]. For By-Product Spec China 1:80–81

Madhavan V, Shilpi A, Anita M, Yoganarasimhan SN (2009) Anti-convulsant activity of aqueous and alcohol extracts of roots and rhizomes of Nymphoides indica (L.) Kuntze in Swiss albino mice. J Nat Remedies 9(1):68–73

Malakar C, Choudhury PPN (2015) Pharmacological potentiality and medicinal uses of Ipomoea aquatica Forsk: a review. Asian J Pharm Clin Res 8(2):60–63

Mei H, Zuo S, Ye L, Wang J, Ma S (2012) Review of the application of the traditional Chinese medicinal herb, Ranunculus sceleratus Linn. J Med Plants Res 6(10):1821–1826

Mithraja MJ, Marimuthu J, Mahesh M, Paul ZM, Jeeva S (2011) Phytochemical studies on Azolla pinnata R. Br., Marsilea minuta L. and Salvinia molesta Mitch. Asian Pac J Trop Biomed 1(1):S26–S29

Ndimele PE, Jimoh AA (2011) Water hyacinth (Eichhornia crassipes (Mart.) Solms.) in phytoremediation of heavy metal polluted water of Ologe lagoon, Lagos, Nigeria. Res J Environ Sci 5(5):424

Oki Y, Une K, Nakagawa K (1989) Relationship between occurrence of aquatic weeds and water quality in the natural water body. Weed Res Jpn 34:97–98

Panda S (2006) Plants of medicinal and ethnobotanical importance in West Bengal: an overview. Medicinal plants: traditional knowledge, 211

Petre JH (1990) U.S. Patent No. 4,951,682. U.S. Patent and Trademark Office, Washington, DC

Pieterse AH (1981) Hydrilla verticillata-a review. Abstr Trop Agric 7(6):9–34

Rahman AHMM, Gulshana MIA (2014) Taxonomy and medicinal uses on Amaranthaceae family of Rajshahi, Bangladesh. Appl Ecol Environ Sci 2(2):54–59

Rai PK, Tripathi BD (2009) Comparative assessment of Azolla pinnata and Vallisneria spiralis in Hg removal from GB pant Sagar of Singrauli industrial region, India. Environ Monit Assess 148(1–4):75–84

Raja MMM, Sethiya NK, Mishra SH (2010) A comprehensive review on Nymphaea stellata: a traditionally used bitter. J Adv Pharm Technol Res 1(3):311

Shendye NV, Gurav SS (2014) Cynodon dactylon: a systemic review of pharmacognosy, phytochemistry and pharmacology. Int J Pharm Pharm Sci 8:7–12

Sivapalan SR (2013) Medicinal uses and pharmacological activities of Cyperus rotundus Linn-a review. Int J Sci Res Publ 3(5):1–8

Srivastava M, Begovic E, Chapman J, Putnam NH, Hellsten U, Kawashima T, Kuo A, Mitros T, Salamov A, Carpenter ML, Signorovitch AY (2008) The Trichoplax genome and the nature of placozoans. Nature 454(7207):955–960

Tungmunnithum D, Pinthong D, Hano C (2018) Flavonoids from Nelumbo nucifera Gaertn., a medicinal plant: uses in traditional medicine, phytochemistry and pharmacological activities. Medicines 5(4):127

Uka UN, Chukwuka KS (2011) Utilization of aquatic macrophytes in Nigerian freshwater ecosystem. J Fish Aquat Sci 6(5):490

Uka UN, Mohammed HA, Ovie SI (2009) Current diversity of aquatic macrophytes in Nigerian freshwater ecosystem. Braz J Aquat Sci Technol 13(2):9–15

Chapter 5
Nutritive Value of Wetland Flora

5.1 Values and Function of Wetland Plants

Wetland flora is of great importance from few viewpoints. Among which one of them is their nutritive value. These are used as one of the main food sources for a variety of animals present on wetland which includes many invertebrates such as insects, and many vertebrates such as beaver, mouse, waterfowls and many muskrat etc. It simply means that these plants have valuable potential as a livestock fodder.

Wetland plants grow abundantly in several water bodies like river, lakes and watercourses throughout the world (Crow and Hellquist 2006). Some of these are emergent, some are floating plants, and some are submerged ones. Wetland flora has been used as food by both animals and humans. Different components of those safe to eat flora are used as a supply of protein and different nutrients. In previous couple of decades' specific emphasis has been given everywhere in the globe to discover all of the viable approaches to apply those flora as an element in animal feeds (Tacon et al. 2009).

In last few years studies have been performed with the intention to make low cost fish feed. All this is done by experimenting aquatic plant based cheaper protein sources in these feed. This precise study is likewise carried out to look for freely obtainable and healthful protein source. One distinguished method of those is making use of plant sources, of each aquatic and terrestrial hydrophyte. Different elements of those plant lives were studied by many researchers to search for alternate cheap sources of nutritive feed for animals (Bairagi et al. 2002, 2004).

Research carried out in ancient times has revealed that these weeds grow in wetlands contain significant amount of minerals and proteins. As those weeds stay unutilized and additionally make aquatic environment damaging for fish culture, those may be transformed into those may be transformed into precious fish flesh through including into fish diets. However numerous constraints are there, low protein contents, amino acid imbalance, presence of anti-dietary factors, extra quantity

M. A. Butt et al., *Wetland Plants*, https://doi.org/10.1007/978-3-030-69258-2_5

of crude fibers in a few plants and occurrence of hemicelluloses, cellulose, and lignin (Kar and Ghosh 2008; Khan and Ghosh 2012). Therefore a few processing approach has been followed to decorate the nutrient cost of those plants' ingredients. Trials have additionally been made in order to raise the bioavailability of these nutrients, to lessen or get rid of anti-dietary elements and crude fibers and additionally to feature lacking flavors required.

Pakistan is home to pretty numerous sorts of wetland vegetation. Public difficulty that aquatic vegetation be eliminated robotically from water as opposed to killing them through herbicides. They are available as feed for livestock in a huge quantity. But their usage as feedstuffs depends upon information of nutrient composition of these plants, as wetland plants differ in their chemical composition depending upon season, species and their location (Eviner 2004).

5.2 General Benefits of Wetland Plants

Wetland plants have plenty of benefits depending upon their growth form.

- Emergents provide significant natural benefits to ponds. As they provide nesting habitat for bird species. Many other aquatic bird species, utilize these plant habitats during migration. Vegetative portion of plants that grows above the surface of water are utilized by mammals as food. Additionally, the seed heads of spike rushes, bulrushes and nut sedges are appealing to many waterfowl species as a meals source. A pond controlled for flora and fauna must try to have plenty of emergent flora.
- In general floating plants like duckweed and water meal offer few advantages to a pond until the pond is precisely a natural pond. The equal species of waterfowl that consume submerged flora may even consume this small floating vegetation. Pond owners who value the natural, aesthetic potentials of a pond need water lilies and their brightly colored flowers. Some fish also like the shade provided by the large leaves of water lilies. The key isn't always to permit the lilies to overrun the pond's shallow areas, preserving approximately 15–20% coverage.
- Submerged flora is vital to a well-dependent fish assemblage. They not only gave safety for small fish from predators but also produce large numbers of plants for other fish populations. This percentage balances the predator-prey relationship between different fish species so that all species have all sizes of fish represented in the group.
- Submerged flora is also a vital meals supply for lots species of waterfowl in the form of plants dwelling invertebrates or the plants themselves. Many species of egrets and herons hunt small fish at the area where submerged plants are densely

grow. A pond controlled for flora and fauna need submerged plants as a habitat component. They directly affect the water quality. Their capacity to place oxygen into the water is a clear involvement, but they also provide for long-term storage of nutrients that might otherwise be used to create nuisance levels of planktonic or filamentous algae. Ponds with beds of submerged plants have less difficulty with algae (Fig. 5.1).

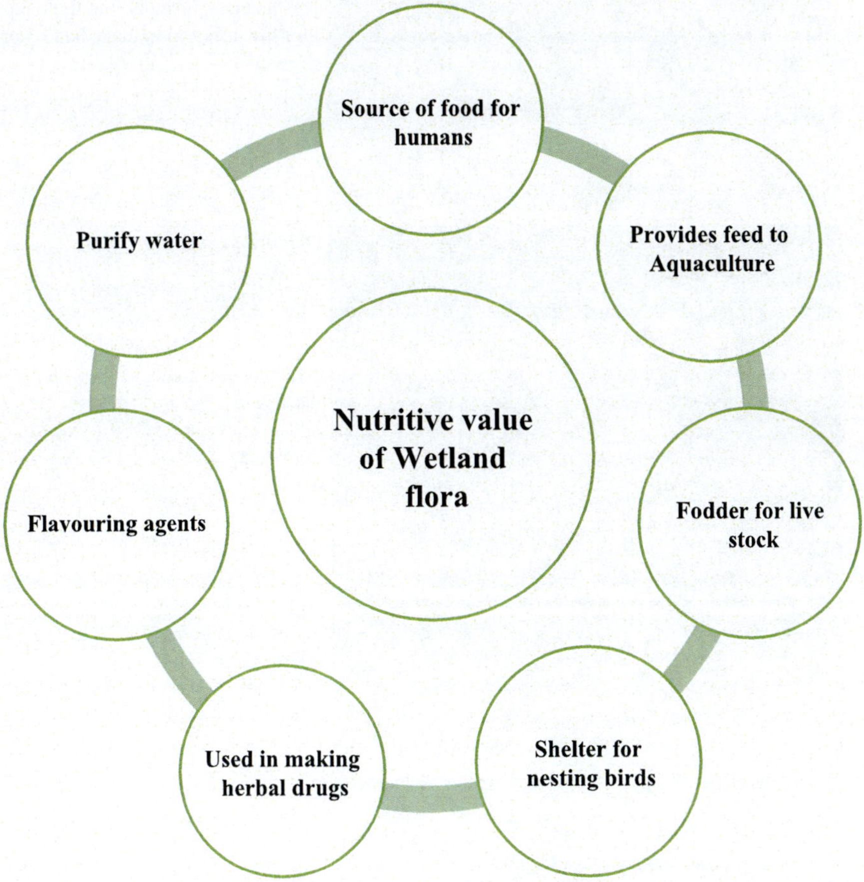

Fig. 5.1 Flow chart of wetland plants nutritive values

5.3 Some Important Nutritious Plants of Wetlands (Table 5.1)

	***Adiantum incisum*** **Forssk. (Pteridaceae)** • Helps to treat coughs, bronchitis, heavy menstruation with cramps. • It is also used to loosen chest congestion. • Some people apply it to the scalp for hair loss and to make hair color darker.
	***Galinsoga parviflora*** **Cav. (Asteraceae)** • Leaves are high in protein so it is cooked as vegetables • Remove skin inflammation by rubbing the affected skin with the leaves. • The leaves are used for wound dressing
	***Centella asiatica*** **(L.) Urb. (Apiaceae)** • Rich in triterpenoids and beneficial carotenoids vitamins B and C proteins. • Flavonoids, volatile oils, tannins, and polyphenol. • Treat diabetes, heal wounds, enhance memory, act as antioxidant, and have neuroprotecting activities

	***Impatiens bicolor* Royle (Balsaminaceae)** • Leaves of this plant has rich amount of antioxidants in it which helps to boost metabolism • Also exhibits nematicidal activity which act as pesticides in agriculture • Also have anti-inflammatory properties due to which it helps sooth the nettle sting area.
	***Polygonum affine* D. Don (Polygonaceae)** • Tannins are present in huge amount which helps to reduce inflammation. • Flavanoids are present in the leaves of this plant which helps to cure indigestion issues. • Extract of root is used in making medicine which treat fever.
	***Leontopodium nivale* subsp. *alpinum* (Cass.) Greuter (Asteraceae)** • This plant is found in higher altitudes so it has metabolites in it which protect the human skin from damage due to ultra violet rays. • Also it shows anti-inflammatory, antimicrobial and anti-parasitic activites.

	***Pedicularis pectinata* (Orobanchaceae)** • Whole plant contains phytochemicals like phenols, flavonoids and alkaloids. • Presence of these compounds possess immunomodulatory, antioxidant, antidiabetic, anti-inflammatory, antibacterial, antitumor, neuroprotective, antipyretic, diuretic and DNA-repairing properties.
	***Lemna minor* L. (Araceae)** • This plant is high in protein content so it can be consumed as perfect supplement to the rice-based staple food. • Aqua culture feed is also prepared from this plant as it contains amino acids and proteins in excessive amount.
	***Schoenoplectus litoralis* (Schrad.) Palla (Cyperaceae)** • Plant contains rich metabolites in it so it is used as fodder for livestock which helps increases milk production in them. • Due to high nutritious stem of this plant is cooked as vegetable or eaten raw.

	***Persicaria hydropiper* (L.) Delarbre (Polygonaceae)** • Leaves of this plant are rich in antioxidants. • Leaves contain rutin, which helps strengthen fragile capillaries and thus helps prevent bleeding.
	***Salvia plebeia* R.Br. (Lamiaceae)** • All parts of this plant are rich in minerals and high in calcium content which is good for bones and teeth. • The root, shoot and seeds contains rich amount of nutrients and mineral which can be used as a food source.
	***Schoenoplectus lacustris* (L.) Palla (Cyperaceae)** • Plant as anti-cancerous properties in it. • Roots act as astringent and diuretic.

***Gnaphalium uliginosum* L. (Asteraceae)**

- This plant is rich in tannins, volatile oil and pro-vitamin A.
- The whole plant is anti-inflammatory, astringent, diaphoretic and diuretic.
- It may also have aphrodisiac and anti-depressant effects.
- Also use to treat high blood pressure.

***Bistorta amplexicalulis* (D.Don) green (Polygonaceae)**

- Root stock of this plant contains tannins in huge amount so the paste is used to cure bleeding wounds and also acts as anti-inflammatory.
- Also taken orally to purify blood.
- Good remedy for ulcer

***Bidens pilosa* L. (Asteraceae)**

- The fresh or dried tender shoots and young leaves are used as a leaf vegetable especially in times of food scarcity.
- It is an ingredient of sauces accompanying the staple food.
- Leaves roots and seed possess anti-dysenteric, antibacterial, antimicrobial, anti-inflammatory, diuretic, antimalarial, hypotensive and hepato protective activities.

Table 5.1 Examples of nutritional contents of some wetland plants

Sr. no.	Plant name	Common name	Nutritive values	References
1.	*Alisma plantago-aquatica* L.	Water plantain	Whole plant is rich in triterpenes which is used as a diuretic and to control kidney and liver problems. It is a good source of dietary fiber, which helps with proper digestion and helps relieve constipation. Presence of calcium helps to strengthen muscles, teeth, and bones. Due to the presence of serum nitroglycerine it helps in lowering blood cholesterol level.	Tinkov et al. (2016)
2.	*Alternanthera philoxeroides* (Mart.) Griseb.	Alligator weed	This plant is a rich source of carbohydrates, proteins, magnesium, calcium, and other minerals. Due to the presence of theses supplements it is suitable green fodder substitute for animals.	Sushil and Kamlesh (2005)
3.	*Ceratophyllum demersum* L.	Hornwort	The plant is a good fodder and rich in protein. Fried plant is also good manure. The extract of this plant contains polysaccharides, including D-xylose, D-galactose, L-rhamnose, L-arabinose, uronic acid, and acidic xylan all these compounds is beneficial for cardiac health.	Naskar (1986)
4.	*Colocasia esculenta* (L.) Schott	Taro	Due to presence of high starch in rhizomes of this plant it is eaten raw by humans. Corm of this plant is highly nutritious and cooked as vegetable. Calcium, phosphorus, and vitamins A and B are found in rhizome of this plant. The leaves and petioles are good sources of calcium, protein, potassium, iron and vitamins A, B, and C so they it can be eaten as a vegetable.	Prajapati et al. (2011), Jianchu et al. (2001)
5.	*Eichhornia crassipes* (Mart.) Solms	Water hyacinth	Leaves of this plant is rich in proteins so that's why it is used as aqua feed, animal feed, fertilizer, water purification, biogas production, even food for human. It also contains high levels of hemicellulose and cellulose, which serve as energy sources for ruminants	Mukherjee and Nandi (2004), Hossain et al. (2015)

(continued)

Table 5.1 (continued)

Sr. no.	Plant name	Common name	Nutritive values	References
6.	*Eleocharis dulcis* (Burm.f.) Trin. ex Hensch.	Chinese water chestnut	In China, it is grown in field as a rotational crop with other aquatic plant species such as rice (*Oryza* spp.), lotus (*Nelumbo* spp.) and arrowhead (*Sagittaria* spp.). The highly required after corms are produced on underground rhizomes. Corms are high in carbohydrates, though low in protein.	Zhan et al. (2014)
7.	*Elodea canadensis* Michx.	Canadian pondweed	Protein content is high in this plant due to which it is used as food by humans. Also used as shelter for aquatic animals. In some countries this plant is used as food for many birds, including coots, ducks, grebes, geese, marsh birds, swans, shore birds and game birds.	Manzoor et al. (2017)
8.	*Ipomoea aquatica* Forssk.	Water spinach	The young leaves are nutritious and are boiled or fried in oil and eaten as a vegetable as they are high in protein content. It also contains some insulin like properties and some research also shows blood sugar lowering effect. This plant has greater tendency to absorb heavy metals from the soil. Older leaves of this plant are high in calcium content so it is cooked as spinach.	Prasad et al. (2008), Ngamsaeng et al. (2004)
9.	*Ludwigia adscendens* (L.) H. Hara	Floating primrose-willow	Rich in phytochemicals which are used to treat skin related issues also juice of this plant is good during ulcer pain.	Ahmed et al. (2005)
10.	*Mentha aquatica* L.	Water mint	The essential oil in the leaves is antiseptic helps in healing wounds. The leaves are antiseptic, anodyne, astringent, and antispasmodic, cholagogue, carminative, emetic, diaphoretic, stimulant, refrigerant, tonic and used in stomachache.	Mulas (2006)
11.	*Monochoria hastata* (L.) Solms	Arrow leaf pondweed	Tender stalk and leaves are eaten as vegetable as they are rich source of phytochemicals in it. Rootstocks are used as food for cattle. Also the fruit of this plant is eaten by aquatic amphibians.	Shankar and Mishra (2012)

(continued)

Table 5.1 (continued)

Sr. no.	Plant name	Common name	Nutritive values	References
12.	*Nelumbo nucifera* Gaertn.	Indian lotus	This plant can be eaten as a vegetable or used in a variety of dishes as it is rich in chemical constituents. The seeds are widely sold in markets. Seeds are used to treat cancer, tissue inflammation, also they are anti-emetic, and are used as a diuretic. The antioxidant properties of have also been well established in the stamens, leaves, seeds and rhizomes. Root of this plant contains vitamins and several minerals so its extract has shown psychopharmacological, diuretic, anti-obesity, anti-diabetic, antipyretic, hypoglycemic, and antioxidant activities. Seeds of this plant contain chemicals like protein, polyphenols, polysaccharides which help to fight against high blood pressure, gallstones and diabetes. Seed having water-soluble polysaccharides also shown to promote lymphocyte transformation thus enhance the immune system functions.	Mehta et al. (2013), Mukherjee et al. (1996)
13.	*Neptunia oleracea* Lour.	Water mimosa	Young plants are cooked and eaten as a green vegetable that is likely high in protein. Leaves of this plant contains huge amount of carotenes in it so medicinally it is used as body detoxifier. Leaves are bandaged on the area of severe allergy which helps to normalize the skin. Dried plant contains vitamin E, vitamin C, xanthophyll, carotenes, phenolics and tannins which are mixed with fodder of animals to increase meat production.	Bhunia and Mondal (2012), Sagolshemcha and Singh (2017)
14.	*Phragmites australis* (Cav.) Trin. ex Steud.	Common reed	Root of this plant contains sugar and it is cooked as potatoes and texture is perfect when root is immature in some countries roots are dried and add it into porridges. Stems of this plant are reported to contain protein, carbohydrates and fiber. The stem is antiemetic, antidote, refrigerant and antipyretic. Seed are highly nutritious and grinded to form powder which is used as flour and helps to boost metabolism.	Rao et al. (2007)

(continued)

Table 5.1 (continued)

Sr. no.	Plant name	Common name	Nutritive values	References
15.	*Rorippa islandica* (Oeder) Borbás	Watercress	Watercress is a source of iron, iodine, vitamin A, B, and C; and is eaten as a fresh salad or cooked as a green vegetable. This plant contains vitamin C in moderate amount so it is good remedy to boost immunity of body and helps cure skin related disorders as well as fungal and bacterial infections. It also contains high amounts of iron, manganese, beta carotene and flavonoids. From the leaves of this plant hair tonic is made and it also helps during hair loss and alopecia. Leaves of this plant are used in salads as a substitute for mustard leaves. Leaves are also used in making green smoothie for weight loss. This plant is a great source of zeaxanthin and lutein which is helpful to fight against macular degeneration. Studies have shown that the leaves also have antibiotic properties.	Garnock-Jones (1978), Klimešová et al. (2004)
16.	*Sagittaria sagittifolia* L.	Arrowhead	Arrowhead has good quantity of phosphorus so consume arrowhead regularly to make our bones strong. Recent research also link the proper consumption of phosphorus rich foods with heart health. Taking an adequate amount of arrowhead will certainly help to lower the risk of developing heart problems like stroke and other heart related diseases since arrowroot consists of considerable amount of potassium. Vitamin B6 contained in arrowhead is required to produce hemoglobin in the blood. Arrowhead is good source of Iron; therefore arrowhead intake is essential to eradicate different reasons of fatigue, which occur in both males and females. In this way consumption of copper rich food like arrowhead helps in defending against graying of the hair. Magnesium contained in arrowhead helps to regulate blood sugar status, therefore encouraging normal blood pressure.	Leung et al. (2013), Li et al. (2007)

(continued)

Table 5.1 (continued)

Sr. no.	Plant name	Common name	Nutritive values	References
17.	*Trapa natans* L.	Water caltrop	The large seeds of this plant are typically roasted, boiled, or consumed raw. Recent evidence suggests that prehistoric civilizations relied on the seeds of water caltrop especially in times of cultivated crop failure. Fruit of this plant is a good source of nutrition having considerable amount of carbohydrate, protein, and vitamins which is used by females to strengthen their muscles after delivery.	Rai and Sinha (2001), Shalabh et al. (2012)
18.	*Typha latifolia* L.	Cattail	Cattail consists of good amount of both soluble and insoluble fiber which is essential for improving the digestion process. Cattail being rich in carbohydrates and calories enables the weight gain process. Adrenal glands are provided the appropriate support to decrease levels of stress by the content of protein and carbohydrates present in Cattail. Cattail consists of good amount of carbohydrate content. It means it has the ability to offer you greater levels of energy.	Yan et al. (2018)
19.	*Vallisneria spiralis* L.	Eel grass	Young leaves eaten as salads. Largely it is used in domestic aquaria and it has great capacity in purifying water, so fish culture is helped directly or indirectly.	Gąbka (2002)
20.	*Zizania palustris* L.	Wild rice	Wild rice provides food and shelter for fish and wildlife, most notably, migratory waterfowl. Today, wild rice is exploited both as a subsistence food and as a cash crop. Compared with other cereals, it is high in protein and low in fat. It is used as remedy for heart ailments, burns, nephrosis, hepatosis, stomach ailments and pulmonosis. The wild populations constitute an excellent fodder, especially for birds.	Kahler et al. (2014), Porter (2019)
21.	*Schoenoplectus californicus* (C.A.Mey.) Soják	Giant bulrush	The young sprouts and shoots can be eaten in salads, the pollen is used as a flavoring, and the roots and unripe flower heads can be boiled as a vegetable.	Rondón et al. (2003), Hidalgo-Cordero and García-Navarro (2017)
22.	*Plantago lanceolata* L.	Plantain	Seeds can be eaten raw or cooked; and they can be ground into flour and used in making diet roti.	Cranston et al. (2015), Grigore et al. (2015)

(continued)

Table 5.1 (continued)

Sr. no.	Plant name	Common name	Nutritive values	References
23.	*Equisetum arvense* L.	Horsetail	Leaves of this plant contains rich amount of proteins so it is used in making herbal shampoos. Ariel part of this plant is edible and cooked as vegetable.	Carneiro et al. (2013)
24.	*Persicaria maculosa* gray	Lady's thumb	The leaves, young shoots, flowers and seeds are edible. Lady's thumb contains natural fibers, sugars, fats and tannins. They also contain high amounts of phenolic acids.	Mosaferi et al. (2011)
25.	*Rumex nepalensis* Spreng.	Nepal dock	Young leaves and shoots are edible in small quantities due to very high oxalic acid content. Many people enjoy preparing the leaves and shoots cooked as a vegetable because once cooked, the oxalic acid is greatly reduced. Roots are edible also when cooked.	Shaikh et al. (2018), Devkota et al. (2015)
26.	*Hydrocharis dubia* (Blume) backer	Frog-bit	Young leaves and young inflorescences can be eaten. Leaves are rich in nutrients.	Bean (2011)

References

Ahmed F, Selim MST, Shilpi JA (2005) Antibacterial activity of Ludwigia adscendens. Fitoterapia 76(5):473–475

Bairagi A, Ghosh KS, Sen SK, Ray AK (2002) Enzyme producing bacterial flora isolated from fish digestive tracts. Aquac Int 10(2):109–121

Bairagi A, Sarkar Ghosh K, Sen SK, Ray AK (2004) Evaluation of the nutritive value of Leucaena leucocephala leaf meal, inoculated with fish intestinal bacteria Bacillus subtilis and Bacillus circulans in formulated diets for rohu, Labeo rohita (Hamilton) fingerlings. Aquac Res 35(5):436–446

Bean AR (2011) Hydrocharis dubia (Blume) backer (Hydrocharitaceae) is an alien species in Australia. Austrobaileya 8(3):435–437

Bhunia D, Mondal AK (2012) Systematic analysis (morphology, anatomy and palynology) of an aquatic medicinal plant water mimosa (Neptunia oleracea Lour.) in eastern India. Int J Life Sci Biotechnol Pharma Res 1(2):290–319

Carneiro DM, Tresvenzol LMF, Jardim PCBV, Cunha LCD (2013) Equisetum arvense: scientific evidences for clinical use. Int J Biol Pharm Allied Sci 2(8):1579–1596

Cranston LM, Kenyon PR, Morris ST, Kemp PD (2015) A review of the use of chicory, plantain, red clover and white clover in a sward mix for increased sheep and beef production. J N Z Grasslands 77:89–94

Crow GE, Hellquist CB (2006) Aquatic and wetland plants of northeastern North America, volume II: a revised and enlarged edition of Norman C. Fassett's a manual of aquatic plants, volume II: angiosperms: monocotyledons, vol 2. University of Wisconson press

Devkota SR, Paudel KR, Sharma K, Baral A, Chhetri SBB, Thapa PPU, Baral KP (2015) Investigation of antioxidant and anti-inflammatory activity of roots of Rumex nepalensis. World J Pharm Pharm Sci 4(3):582–594

Eviner VT (2004) Plant traits that influence ecosystem processes vary independently among species. Ecology 85(8):2215–2229

Gąbka M (2002) Vallisneria spiralis (Hydrocharitaceae)-a new species to the Polish flora. Fragmenta Floristica et Geobotanica, Series Polonica 9:67–73
Garnock-Jones PJ (1978) Rorippa (Cruciferae, Arabideae) in New Zealand. NZ J Bot 16(1):119–122
Grigore A, Bubueanu C, Pirvu L, Ionita L, Toba G (2015) Plantago lanceolata L. crops-source of valuable raw material for various industrial applications. Sci Paper Ser A Agron 58:207–214
Hidalgo-Cordero JF, García-Navarro J (2017) Review on the traditional uses and potential of totora (Schoenoplectus californicus) as construction material. Mater Sci Eng 245:022068
Hossain ME, Sikder H, Kabir MH, Sarma SM (2015) Nutritive value of water hyacinth (Eichhornia crassipes). Online J Anim Feed Res 5:40–44
Jianchu X, Yongping Y, Yingdong P, Ayad WG, Eyzaguirre PB (2001) Genetic diversity in taro (Colocasia esculenta Schott, Araceae) in China: an ethnobotanical and genetic approach. Econ Bot 55(1):14–31
Kahler AL, Kern AJ, Porter RA, Phillips RL (2014) Maintaining food value of wild rice (Zizania palustris L.) using comparative genomics. In: Genomics of plant genetic resources. Springer, Dordrecht, pp 233–248
Kar N, Ghosh K (2008) Enzyme producing bacteria in the gastrointestinal tracts of Labeo rohita (Hamilton) and Channa punctatus (Bloch). Turk J Fish Aquat Sci 8(1):115–120
Khan A, Ghosh K (2012) Characterization and identification of gut-associated phytase-producing bacteria in some freshwater fish cultured in ponds. Acta Ichthyol Piscat 42(1):37
Klimešová J, Martínková J, Kočvarová M (2004) Biological flora of Central Europe: Rorippa palustris (L.) Besse. Flora Morphol Distrib Funct Ecol Plants 199(6):453–463
Leung HH, Fang EF, Ng TB (2013) A landscape of the health benefits of different natural protease inhibitors. In: Antitumor potential and other emerging medicinal properties of natural compounds. Springer, Dordrecht, pp 213–224
Li L, da Silva JAT, Cao B (2007) Aquatic vegetable production and research in China. Asian Australas J Plant Sci Biotechnol 1(2):37–42
Manzoor MF, Ahmad N, Manzoor A, Kalsoom A (2017) Food based phytochemical luteolin their derivatives, sources and medicinal benefits. Int J Agric Life Sci IJAL 3(11):1
Mehta NR, Patel EP, Patani PV, Shah B (2013) Nelumbo nucifera (Lotus): a review on ethanobotany, phytochemistry and pharmacology. Indian J Pharm Biol Res 1(4):152–167
Mosaferi S, Keshavarzi M, Ghadam P (2011) Biosystematic study of annual species of Persicaria from Iran using SDS-PAGE. Phytol Balcan 17(2):185–190
Mukherjee R, Nandi B (2004) Improvement of in vitro digestibility through biological treatment of water hyacinth biomass by two Pleurotus species. Int Biodeterior Biodegradation 53(1):7–12
Mukherjee PK, Balasubramanian R, Saha K, Saha BP, Pal M (1996) A review on Nelumbo nucifera gaertn. Anc Sci Life 15(4):268
Mulas M (2006) Traditional uses of Labiatae in the Mediterranean area. In: I International Symposium on the Labiatae: advances in production, biotechnology and utilisation 723, pp 25–32
Naskar KR (1986) Recent trends of aquatic weeds management through proper utilization. Plant Sci Res. vol 1:188–193
Ngamsaeng A, Thy S, Preston TR (2004) Duckweed (Lemna minor) and water spinach (Ipomoea aquatica) as protein supplements for ducks fed broken rice as the basal diet. Livest Res Rural Dev 16:18–24
Porter R (2019) Wildrice (Zizania L.) in North America: genetic resources, conservation, and use. In: North American crop wild relatives, vol 2. Springer, Cham, pp 83–97
Prajapati R, Kalariya M, Umbarkar R, Parmar S, Sheth N (2011) Colocasia esculenta: a potent indigenous plant. Int J Nutr Pharmacol Neurol Dis 1(2):90
Prasad KN, Shivamurthy GR, Aradhya SM (2008) Ipomoea aquatica, an underutilized green leafy vegetable: a review. Int J Bot 4(1):123–129
Rai UN, Sinha S (2001) Distribution of metals in aquatic edible plants: Trapa natans (Roxb.) Makino and Ipomoea aquatica Forsk. Environ Monit Assess 70(3):241–252

Rao AN, Johnson DE, Sivaprasad B, Ladha JK, Mortimer AM (2007) Weed management in direct-seeded rice. Adv Agron 93:153–255

Rondón XJ, Banack SA, Diaz-Huamanchumo W (2003) Ethnobotanical investigation of caballitos (Schoenoplectus californicus: Cyperaceae) in Huanchaco, Peru. Econ Bot 57(1):35

Sagolshemcha R, Singh R (2017) Traditional and biological uses of Neptunia oleracea lour: an overview. Int J Curr Res 9(06):51689–51694

Shaikh S, Shriram V, Srivastav A, Barve P, Kumar V (2018) A critical review on Nepal dock (Rumex nepalensis): a tropical herb with immense medicinal importance. Asian Pac J Trop Med 11(7):405

Shalabh B, Akash J, Jasmine C (2012) Trapa natans (water chestnut): an overview. Int Res J Pharm 3(6):31–33

Shankar LH, Mishra PK (2012) Study of aquatic medicinal plants of Hazaribag district of Jharkhand, India. Int Res J Pharm 3(4):405–409

Sushil K, Kamlesh V (2005) Nutritive value of alligator weed [Alternanthera philoxeroides (Mart.) Griseb.] and its possible utility as a fodder in India. Indian J Weed Sci 37(1/2):152

Tacon AG, Metian M, Hasan MR (2009) Feed ingredients and fertilizers for farmed aquatic animals: sources and composition, vol 540. Food and Agriculture Organization of the United Nations (FAO), Rome

Tinkov AA, Nemereshina ON, Suliburska J, Gatiatulina ER, Regula J, Nikonorov AA, Skalny AV (2016) Comparative analysis of the trace element content of the leaves and roots of three Plantago species. Biol Trace Elem Res 173(1):225–230

Yan N, Du Y, Liu X, Chu C, Shi J, Zhang H, Liu Y, Zhang Z (2018) Morphological characteristics, nutrients, and bioactive compounds of Zizania latifolia, and health benefits of its seeds. Molecules 23(7):1561

Zhan G, Pan LQ, Mao SB, Zhang W, Wei YY, Tu K (2014) Study on antibacterial properties and major bioactive constituents of Chinese water chestnut (Eleocharis dulcis) peels extracts/fractions. Eur Food Res Technol 238(5):789–796

Chapter 6
Medicinally Important Wetland Flora

In traditional medicine system traditional knowledge has been utilized for treating or curing several ailments and therapies. The use of traditional system of medicine is commonly practiced in most of the countries throughout the world apart from development in modern medicine system. This is probably due to a variety of complex socio-financial factors, lack of contemporary facilities, education, expensive treatments and beliefs. The roots of traditional medicinal systems include the use of different plants for healing purpose with transfer of knowledge from one generation to other.

The traits in current medicinal structures originate from these in one manner or another. Fast advances in science and knowledge related to chromatography, pharmacognosy, molecular biology and plant biotechnology, has extremely increased the principles of the traditional medicinal systems that describe active organic compounds and their uses (Kar 2003). Usage of alternative medicines are denoted to as practices which involves medicinal therapies and byproducts employed for the benefit of human having knowledge of their toxic, harmful and beneficial effects (Dasgupta and Hammett-Stabler 2011). The use of alternative drug treatments is primarily based on religion, belief, social factors superstitions and family-based totally medicinal structures mostly no longer recognized by the present day clinical and biological studies (Sampson 1995). These medications are used for treating prolonged diseases conditions like headache, fever etc. and serious diseases like cancer (Peregoy et al. 2014). Alternative and complementary medicines encompass a big range of health care practices that aren't considered to the lifestyle without integration into the fitness system of the country (WHO 2000). However, alternative medicinal systems are being used as complementary medicine (CM) or integrative medicine (IM) used with functional medicines (Zeller et al. 2013), CAM (complementary and alternative medicines) (Cassileth and Deng 2004) or TM (traditional medicines) (Ansari and Inamdar 2010).

Medicinal plants are those plants which have beneficial therapeutic and pharmacological effects on health of human beings. These plants are rich in secondary

M. A. Butt et al., *Wetland Plants*, https://doi.org/10.1007/978-3-030-69258-2_6

metabolites like sterols, saponins, alkaloids, volatile oils; flavonoids etc. and they are utilized since ancient times by humans (Baytop 1999). Previous literature shows use of plants as medicine by Egyptian papyrus and Chinese scriptures followed by the improvement of other traditional medicinal systems. Well known traditional system of medication involves; Techiman-Bono ethno-medical system, Western Orthodox, naturopathy, Native American medicine system, reiki, Ayurvedic, Dormaa medicine system, Tongan, Greco-Arab, Chinese, Unani, African systems and Iranian system etc. (Warren and Green 1982; WHO 2001; Shankar and Liao 2004; Isola 2013). In Chinese traditional system of medicine it involves about 5000 medicinal plants (Li 2000). Wetland atmosphere has played a widespread function in human civilization on the grounds that a long time ago. Generally people desired to stay close to water bodies or wetland systems in historic times; where they may develop and irrigate plants for food, garments and abode. A have a look at of all extremely good civilizations like Indus valley, Mesopotamian and Nile civilizations and modern-day Roman, Greek and Inca civilizations also prove that human beings continually favored to live close to rivers, lakes and oceans to get sustenance for food, shelter and residence. Aquatic flora have important role in human development (Capers et al. 2010) by providing manure, food, source of medicines and fiber.

Aquatic flora is considered as primary producers of the aquatic ecosystem as they convert light energy into chemical energy. In addition to this, aquatic flora provides food and shelter for aquatic fauna like insects, fish and other aquatic amphibians and reptiles (Andreasen and Stryhn 2008). They also used to control phytoplankton and mosses from water (Oyedeji and Abowei 2012). They firmly anchor the sediments of soft-bottom habitats. The diversity and fitness of aquatic ecosystem depends on the population of hydrophytes in it (Flint and Madsen 1995). The importance of aquatic ecosystem increases as the hydrophytes are used for phytoremediation of heavy metals and for biomonitoring. Demand of aquatic plants increases as it is use in aquarium industry in several countries like United States (Crosson 2010), Australia (Petroeschevsky and Champion 2008) and Europe (Brunel 2009). Aquatic flora is rich in secondary compounds that are commercially utilized for treating several disorders and diseases. Climate of wetlands play an important role to the ecosystem of area, species richness and habitat (Kirim et al. 2014). Aquatic plant life is primary constituent of aquatic ecosystem with unique ecological area of interest and unique adaptations. They grow as emergents, submerged or as floaters (Dodds 2002). Aerenchyma is the most common type of adaptation of aquatic plants. Also, amphibian vegetation have adapted to grow both on soils absolutely saturated with water and below aquatic situations and are also taken into consideration as wetland or aquatic flora (Keddy 2010).

According to an estimate about 50,000–80,000 flowering plants are used as medicine throughout the world by continuous collection or domestication from the wild. These medicinal plants can be used as supportive medicine, synergic medicine, or preventive medicine (Rasool Hassan 2012). They provide raw material for synthesizing herbal drugs like biocides, laxatives etc. In last few years the demand of these plants have been increased resulting in increase in cultivation of these medicinal species, screening of new potential medicinal plants or phytochemicals, alteration of phytochemicals into registered herbal drugs and eventually rise in trade.

However, the evaluation process of isolation of phytochemicals and their pharmacological activities needs time and multidisciplinary approaches. Comparison of research on aquatic plants has found out that this flora is the most disregarded or underutilized flora. Many among them have high potential to be used as medicinal plants, but their usages are limited to particular zones of the world (Swapna et al. 2011), mainly because of their negligence or likes or dislikes or eating conducts of people. More than 50% of people in many parts of world live close to the water habitat use aquatic plants as medicine, food and ornamental in religious offerings (Munasinghe et al. 2010).

6.1 Major Aquatic Medicinal Plants

6.1.1 **Caltha palustris** ***var.*** **alba** ***(Cambess.) Hook.f. & Thomson (Ranunculaceae)***

Dried leaves are crushed in powder which is then taken to cure antispasmodic patients. Fresh leaves are chewed to relieve from painful menstruation and also act as diuretic. Aqueous extract of shoots and root is used in cough. Decoction made from dried roots is used during urine infections. Fresh flowers are boiled in water to make syrup and taken to sooth stomach inflammation (Kumar et al. 2009). Half teaspoon of dried whole plant is infused in water for 5 min and after cooling this infusion is rubbed on burned skin to relief from pain. Juice formed from mixing fresh flowers and leaves helps in quick healing of wounds and also effective in removing warts. Decoction made from crushed dry roots was taken 1 table spoon 3 times a day to boost metabolism and cure anemia (Martynov et al. 2017). Decoction made from dried leaves have weak anti-tumor properties (Mubashir et al. 2014). Infusion made by whole plant dipped in boiled water for overnight and taken 2 times a day to cure whooping cough and menstrual cramps. Decoction of whole plant is good in treating metabolic disorders, diathesis, common colds and act as an anti-cingulate agent (Liakh et al. 2020). Paste made from crushing fresh leaves with water is used to treat skin problems like sunburns, fungal infections.

6.1.2 Cardamine hirsuta *L. (Brassicaceae)*

A part from medicinal uses this plant is edible and used as salads. Juice of whole plant is used in treating oral disorders. Decoction made from whole plant helps to keep immune system strong and it has ability to fight against cancer. This plant is highly medicinal that has proven diuretic and antipyretic properties which plays important role in treating bladder and uraninary inflammations, diarrhea, dysentery and white vaginal discharges (Xie 1996). Decoction made from leaves is taken at night time to cure insomnia and gargling of decoction helps during toothache (Thong and Madhialagan 2014). Paste from fresh leaves is applied on wounds and minor injuries (Jain et al. 2011). Infusion is taken on daily basis for 2 weeks to cure, arthritis, chickenpox, stomach ache and menstrual irregularities (Santhosh and Satyanarain 2010) and as an antispasmodic, analgesic (Manosroi et al. 2016) and diuretic agent. Also, vermifuge, antipyretic, anodyne, antisplasmodic, antiseptic, hypoglycaemic, sedative, hypotensive (Duke and Ayensu 1985), anti-cancer (Chiang et al. 2004), anthelmintic (Manandhar 1986), anti-ulcer (Chung et al. 2011), anti-mutagenic (Chen et al. 2011), anti-allergic (Chen et al. 2012) and anti-inflammatory (Seo et al. 2000) properties also increased the demand of this plant. Poultice of leaves is applied on the area of scorpion bite to relief pain also it is good in treating pain due to insect bites.

6.1.3 **Cyperus alternifolius** *L. (Cyperaceae)*

Dried stems of this plant are used in making bottom of wooden chairs. Pulp of this plant is processed in industries and helps in making paper. Decoction made from boiling whole plant in water is used to treat digestive disorder and also helps in curing different types of fever and to disinfect wounds from germs ash of this plant is applied as a poultice on them.

Powder of dried leaves is taken for few days to treat rheumatoid and arthritis pain. Decoction of leaves is taken as a tonic. Dried leaves and stems are burnt to produce smoke, this smoke act as insect repellent, leaves are also good in sprain (Mollik et al. 2010a, b).

In some countries like Kenya this plant is used to make a charm that is used to induce extra body strength. Also in some cultures root of this plant is used to treat spiritual sickness (Pakia et al. 2003). Decoction made from boiling roots in water helps in curing mouth boils and rushes (Kaigongi and Musila 2015).

Decoction of whole plant is used to treat tuberculosis (Chaachouay et al. 2019). Decoction of root is taken orally to treat kidney disorders specifically for curing kidney stones (Ramli et al. 2015). Juice of leaves is taken empty stomach to treat cardiovascular or blood related problems (Sharma et al. 2017). Root and leaves of this plant also exhibits antifungal, antibacterial antimicrobial, antioxidant and hepato-protective properties.

6.1.4 Persicaria nepalensis *(Meisn.) Miyabe. (Polygonaceae)*

This plant is helpful in curing piles and spermatorrhoea by taking decoction of roots in empty stomach. Whole plant is boiled in water to cure rheumatic pains. Powder of whole plant is mixed with flour to form poultice and it applied on the area to relief muscular pain (Tonny et al. 2017).

Decoction of whole plant is used to treat colic pain, skin disorders such as boils, scabies, ringworms, abscesses, inflammatory conditions like muscular pain, rheumatic pain, knee pain, menstrual pain, gout and amenorrhoea,. In traditional medicine this plant is used to treat dyspepsia, diarrhea, itchy skin and piles (Keya and Rahman 2017). Seeds are called to have purgative, antiemetic and stimulant properties and can be used to get rid of colic pain. Extract obtained from shoots are used to wash the wounds caused by ulcer.

Juice made by mixing fresh plant in water is used as diuretic, anthelmintic, carminative and also can be used for itchiness (Ahmed et al. 2011). Infusion of roots is good tonic and can be used in ulcerative colitis and also good in curing fever (Uddin et al. 2014). To cure stomach disorders juice of leaves is the best remedy.

Extract of this plant has antibacterial, antifungal and insecticidal properties. Root decoction is used orally to treat high fever and vomiting as a home remedy (Sathyavathi and Janardhanan 2011).

6.1.5 **Rorippa islandica** ***(Oeder) Borbás (Brassicaceae)***

Decoction of leaves are used in expelling phlegm during coughing, also helps in curing pneumonia and common colds. Powder of leaves is also used in purifying blood and treating jaundice and liver related problems. The leaves are antiscorbutic (Lassak and McCarthy 2011).

This plant acts as deobstruent, diuretic, cures scurvy, stimulant and used as tonic. Plant is eaten raw to cure kidney disorders, heart related issues and tuberculosis. Paste of plant is applied on the skin to get rid of pimples and acne marks. Decoction of leaves is good remedy for solving indigestion issues also boot up the metabolism. Drinking the decoction of roots can cause sterility or sometimes it also leads to abortion in females (Vardhana 2008).

Decoction of whole plant is used to treat cough, common cold, dyspnea and pyrexia, severe cases of measles with insufficient epidemics, rheumatic pain, muscular swelling pain, sore throat, furunculosis and anthracia, dermatitis, menostasia, severe wounds, edema and jaundice. This plant also helps in expelling phlegm for suppressing cough, relieving exterior syndrome and dispersing cold, promotes circulation of blood and detoxifying liver, disinhibiting succulence and eliminating jaundice (Tynsong et al. 2006).

6.1.6 Veronica anagallis-aquatica *L. (Plantaginaceae)*

Decoction of leaves is also good in purifying blood. Leaves are also used in salads with combination with lemon and vinegar enhances the taste of salads. The root and the leaves are alterative, appetizer and diuretic (Riley et al. 2012).

Juice extracted from the leaves is used in the treatment of scurvy and also purify the blood. Poultice of leaves are applied on cuts, wounds, burns, ulcers, whitlows to relive pain. Decoction of whole plant is taken to regulate menstrual flow. Powder of roots is use to bring vomiting and helps in blood purification which results in quick healing of internal wounds. Root decoction helps in treating tuberculosis. Roots also act as strong laxative and emetic (Bojnanský and Fargašová 2007). Liver disorders and bile flow diseases is also cured by the use of this plant. Different extracts are made mixing the powder of this plant in combination of other plants to cure constipation, indigestion linked with liver disorders and cholecystitis.

It is also used to treat pharyngitis influenza and larynx. The extract of whole plant is cooked in milk which is used during stomach aches, and when this extract is cooked in water then it helps to relieve rheumatic pains (Duke 2002). Fruit of this plant is eaten raw to cure low back pain and paste of roots is applied on bruises to stop bleeding and relieve from swelling and pain. Leaves have sap which is rub on the area where insect bite and helps in lowering the pain.

6.1.7 **Veronica beccabunga** ***L. (Plantaginaceae)***

Juice extracted from the stems of this plant is taken 2 times a day for diuretic patients. This juice is also good for stomach disorders and anemia. Leaves of the plant are used as poultice on wounds for quick healing.

Juice of leaves is used in treating respiratory disorders as cough syrup, in GIT disorders, liver disorders, cystitis, urinary tract and kidney infections, as well as nervous stimulation and gout. Decoction of whole plant is effective remedy in curing inflammation of the mucous membranes of the body, bleeding wounds, skin diseases such as eczema and extreme perspiring feet.

Infusion made from whole plant is effective to cure constipation, reduced urine output, liver dysfunction, pulmonary related diseases and oral disorders such as bleeding gums (Di Giorgio et al. 2008). Powder helps in dissipating phlegm and reduces swelling, (Xue et al. 2019).

Whole plant is used to ease painful urination, to treat respiratory ailments, for loss of appetite, dysentery and hemorrhoids. In some regions of world leaves are rubbed on the surface of wounds to relief from pain (Guarrera and Savo 2016).

6.1.8 Adiantum capillus-veneris *L. (Pteridaceae)*

Powder of dried fronds are mixed with honey and used to cure catarrh infections. Decoction of fronds is also good remedy for sore throats. Dry fronds are also used to clean kitchen utensils. In Unani medicine this plant is used to treat inflammatory diseases (Haider et al. 2011). From this plant an important drug is formed which is used to treat patients of urolithiasis and is used for litholytic activity. Decoction of whole plant is used for its lithotriptic effect and also considered to be capable of releasing stones from bladder and kidney (Ibn Sina 2007; Ghani and Advia 2011).

Decoction of fronds is taken orally to cure respiratory disorders; asthma, dyspnea, coryza and chest pain (Ibn Sina 2005). In earlier days some physicians use this decoction as eye drop to control the fistula lacrymalis condition. Powder is taken orally with water to treat GIT disorders such as diarrhea, dysentery, abdominal cramps and jaundice (Shirazi 1992). It is also reported as headache-preventing agent. Good remedy for dissolving kidney calculi and act as a diuretic agent when orally taken in the form of decoction or infusion. Decoction of fronds taken orally to cure female genital disorders such as amenorrhea. Perfect hair tonic is made from this plant which cures alopecia and enhances hair growth and remove dandruffs from hair (Khodaie et al. 2015).

6.1.9 Epilobium laxum *Royle (Onagraceae)*

Powder of dried leaves is taken with water 2 times a day for curing burning sensation in stomach. Powder of leaves is taken with milk empty stomach helps in curing rheumatic pains. Paste made from the leaves with the help of water and used on snake bite helps relieving pain and also release poison from the body. Whole plant juice is good remedy to treat ulcers, oral infections, and respiratory issues, from oropharyngeal to bronchial, larynx issues, the nasal mucosa, the sinuses and vocal organs. Powder of this plant mixed with other herbs is used as a remedy for benign prostatic hypertrophy, diseases of urinary tract and kidney infections (Ganie et al. 2020).

It is used in the form of herbal tea in many countries and considers it useful for its soothing and astringent properties, also using it as healing tonic. This tea is good remedy for treating hoarseness, pharyngitis, laryngitis, and naso-pharyngitis. As it has expectorant properties due this it helps in thinning of phlegm.

Decoction of whole plant is good diarrhea (Asim et al. 2016). Paste of leaves is applied on the skin to treat dermatosis or it is also used to cleanse the skin. In case of irregular menstruation decoction of whole plant is most effective (Ballabh et al. 2017).

6.1.10 Geranium nepalense *Sweet (Geraniaceae)*

Decoction made from the roots is used to treat ulcer and also use to cure toothache and eye diseases. Poultice is used for the treatment of joint pain. Juice of the plant is used to stop bleeding. Rhizomes are dried and ground into powder form which is mixed with wheat flour. Juice extracted by mixing whole plant in water is good treatment for renal disorders. Whole plant powder act as astringent and also it is antibacterial. In some countries some people use decoction of leaves to cure nervous disorders, paralysis of limbs, joint pains and rheumatism.

Whole plant is good remedy for dysentery, flu, cold, piles and indigestion leads to stomachache. This plant acts as astringent, styptic, used in kidney diseases, dysentery, external and internal bleeding.

Paste of leaves is used to treat skin ulcers and hemorrhoids. Powder made from grinding dry roots mix with milk and taken orally to treat back pain, helps in strengthening bones and muscles and cure gout (Qureshi et al. 2009). In root decoction of this plant pods powder of *Pistacia chinensis* is mixed and used to treat sever cough, fever and urinary complaints. It also used for headache, respiratory diseases, sexual and genital diseases. Decoction of whole plant is used to treat female gynecological disorders including like gonorrhea, leucorrhoea and diabetes.juice extracted from leaves of this plant helps in curing liver related problems specifically jaundice and hepatitis (Shaheen et al. 2017)

6.1.11 Polygonum aviculare *L. (Polygonaceae)*

Different colored dyes are obtained from the whole plant. Whole plant powder is used as flavorings agents in pancakes and biscuits. Decoction of leaves is used to remove kidney stones. Juice of whole plant is taken to cure respiratory disorders like bronchitis and severe cough (Zabihullah et al. 2006). Whole plant infusion is good remedy to treat hemorrhages, piles and severe diarrhea.

Decoction of leaves is taken three times a day helps in excreting stones in urinary bladder and also cure urinary tract disorders. Whole plant taken orally in raw form helps to purify blood. This plant helps to regulate blood circulation in the body. Decoction of roots helps to boost the immune system of the body and protect the body against any kind of infections (Gan et al. 2010).

Paste of leaves is applied on the skin to prevent early aging, wrinkles and freckles as it maintain skin elasticity. It acts as natural moisturizer to the skin and easily removes all the toxins from skin. This plant act as antibiotic as it quickly stops bleeding of wounds, reduces swelling and inflammation. Also remove infection and sepsis from wounds (Narasimhulu and Mohamed 2014).

During sore throat leaves of this plant are eaten to relief pain also cures symptoms like flu, neck stiffness, earache, chest pain and tonsillitis (Mollik et al. 2010a, b).

6.1.12 Nelumbo nucifera *Gaertn. (Nelumbonaceae)*

The root of the plant is diuretic. This plant is used in traditional medicine in many parts of the world. It is used to treat diarrhea, sunstroke, hemorrhoids, dysentery, vomiting of blood, dizziness, bleeding uterus, remove wrinkles and freckles, lower burning sensations, fight against infections, hypertension, cough, urinary disorders and fever. Extract of rhizome have shown psychopharmacological, diuretic, anti-obesity, anti-diabetic, antipyretic, hypoglycemic and antioxidant activities.

Extract made from flowers and floral buds is used in curing diseases like cancer, hypertension, body heat imbalance, weakness, and kidney disorders, sexual disorders in males, stop bleeding wounds, syphilis and to eliminate the toxins from the blood. Seeds of this plant are used as spleen tonic and powder is used against sever cough. Plumule from the ripe seed is used for the treatment of insomnia, nervous diseases, high fevers with hypertension and restlessness.

Stamens of this plant is mixed with milk and taken orally to treat sexual debility in males. Rhizome extract is taken on empty stomach to cure diabetes. Leaves juice is good during weight loss. Paste of leaves are applied of bleeding wounds to stop bleeding.

6.1.13 Plantago lanceolata *L. (Plantaginaceae)*

Dried seeds and the dried husk obtained from this plant are demulcent, emollient and laxative. It is also used to treat dysentery, urinary tract disorders and swollen intestinal canal. The seeds of this plant also help in easing the stool passage. It is also used to treat constipation, dysentery and other intestinal grumbles, having a soothing effect on the digestive system. Also cures the inflammation caused by hemorrhoids. The thick jelly produced by the husk when soak in water is used to absorb toxins of the intestines. It is also effective in reducing cholesterol levels in the blood stream. Oil and paste produced from the parts of this plant is use to solve the skin problems, headaches and irritation on skin. Powder is helpful in maintaining the normal body temperature. It is very effective in dry cough and urine related problems it is also useful in destroying the infection in the body. It also supports the body resistance (Calabozo et al. 2001).

In some countries tincture made from the mashed leaves mixed with alcohol is applied to aching teeth caused by caries. The crushed leaves are applied as a poultice on wounds to stop bleeding. A leaf decoction or infusion is used to wash the eyes to treat eye infection. A tea made from the whole plant is taken against nausea and is used as a mouth wash for aphthae (Guil-Guerrero 2001).

6.1.14 Cyperus glomeratus *L.(Cyperaceae)*

Decoction of leaves is commonly used as astringent, anti-rheumatic, diuretic, styptic, vermifuge and tonic. Infusion is made that is used for the treatment of diarrhea, gastro intestinal disorders, gonorrhea and bleeding piles. Also used to treat tonsils inflammation and throat infections. Roots are boiled to make tea that is used for the gynecological diseases. Also used for kidney problems, diabetes, fevers, bronchitis and cough.

Decoction is used against fungal diseases, skin infection. Use in chronic fevers, abdominal pain and nerve tonic. Also use for blood purification.

Fresh shoots are used as laxatives and anti-diabetic. Fresh leaves are used for fever. Leaves are crushed and use for boils. Decoction is prepared and used for cough. Aqueous leaf extracts are used to reduce glucose level in blood (Table 6.1).

Table 6.1 Medicinal uses of some important wetland medicinal plants

Sr. No	Botanical name/ family	Habit	Plant part used	Medicinal uses	References
1	*Achillea millefolium* Asteraceae	Emergent herb	Whole plant	3 g per day as a tea or infusion is taken to cure skin diseases especially eczema. Dried herb is used 2 to 4g as capsules three times a day in case of painful urination.	Applequist and Moerman (2011)
2	*Ammannia baccifera* Lythraceae	Emergent herb	Leaves	Leaf paste is applied on ringworm. Leaf decoction along with fresh ginger juice is given to cure internal fever.	Loganayaki et al. (2012) and Al-Snafi (2015)
3	*Arundo donax* Poaceae	Emergent herb	Leaves and stem	Paste of leaves is used to treat skin diseases. Decoction of stem is used in treating skin disorders	Ikram et al. (2014)
4	*Bistorta amplexicaulis* Polygonaceae	Emergent herb	Leaves, rhizome, root	Tea made from the leaves of this plant is taken 2 times a day to cure flue and cough. Paste of rhizome is applied on wounds and sores for quick healing. Root powder is taken with milk to check excessive bleeding during menstruation. Decoction of roots is taken 3 times a day to purify blood.	Hamayun et al. (2006)
5	*Centella asiatica* (Linn.) Apiaceae	Emergent herb	Whole plant	It is employed to treat dysentery, skin diseases, brain disorders, tuberculosis and ulcer.	Sardar (2008) and Roy et al. (2013)
6	*Cerastium glomeratum* Caryophyllaceae	Emergent herb	Leaves	Juice of leaves dropped into nostrils to stop nose bleed. The juice of the plant is applied to the forehead to relieve headaches	Chandra and Rawat (2015)
7	*Cyathocline purpurea* Asteraceae	Emergent herb	Roots, leaves	It has antimicrobial, anthelmintic and hypotensive properties including cancers. The roots of this plant are reportedly used to relieve stomach pains.	Parrotta (2001) and Ma et al. (2009)

(continued)

Table 6.1 (continued)

Sr. No	Botanical name/ family	Habit	Plant part used	Medicinal uses	References
8	*Cynoglossum lanceolatum* Boraginaceae	Emergent herb	Leaves, root	Decoction of the leaves is drunk during delivery to promote quick removal of the placenta. The vapor of the crushed leaves is inhaled to treat fever and influenza. The crushed leaves are rubbed on scorpion sting wounds. A decoction made from pounded and soaked roots is drunk by pregnant women to relieve abdominal pain.	Joshi (2016)
9	*Epilobium laxum* Onagraceae	Emergent herb	Leaves, roots	Powder of leaves is taken with milk empty stomach helps in curing rheumatic pains. Paste made from the leaves with the help of water and used on snake bite helps relieving pain and also release poison from the body. Juice obtained from different part of the plant including roots and leaves are taken 3 times a day for the treatment of fever and inflammation. Paste of the leaves also applied on the skin to relief from skin allergies.	Dad and Khan (2013)
10	*Erigeron canadensis* Asteraceae	Emergent herb	Leaves	Infusion made by soaking dried leaves in 500 ml water and taken empty stomach to treat gastrointestinal disorders. Tea of this plant is used to treat gynecological disorders.	Mahmood et al. (2011)
11	*Ipomoea aquatica* Convolvulaceae	Emergent herb	Leaves, latex	Extract made from leaves are used to heal wounds. Paste of leaves and latex is applied on the skin to cure skin allergies.	Malakar and Choudhury (2015)
12	*Lepidium didymum* Brassicaceae	Emergent herb	Leaves	Infusion of leaves is taken 1 liter daily to cure diabetes. Paste of leaves is applied on skin allergies to cure them. Infusion of leaves is good remedy for treating respiratory disorders.	Akhtar et al. (2017)

(continued)

Table 6.1 (continued)

Sr. No	Botanical name/ family	Habit	Plant part used	Medicinal uses	References
13	*Mentha longifolia* Lamiaceae	Emergent herb	Leaves	A tea made from dried leaves is taken 1 cup in morning to cure fevers, headaches. Decoction of fresh leaves is taken 2 cups a day to treat imbalance hormones in females. Leaves of this plant is boiled in milk and taken during abdominal pain.	Mikaili et al. (2013)
14	*Oenothera rosea* Onagraceae	Emergent herb	Leaves	Infusion of leaves is taken 1 cup on empty stomach to treat renal infections. Leaves are chewed in raw form to treat stomach disorders.	Sumitra et al. (2012)
15	*Persicaria barbata* Polygonaceae	Emergent herb	Seeds, root	Seed powder is used 3 times a day to treat diarrhea and dysentery. Paste made from the root of this plant is used to treat scabies.	Tonny et al. (2017)
16	*Persicaria glabra* Polygonaceae	Emergent herb	Whole plant, leaves	The juice of the herb is taken 2 times a day to use as a remedy for fever. Infusion of leaves is used to relieve colic pain.	Joshi and Joshi (2006)
17	*Persicaria maculosa* Polygonaceae	Emergent herb	Whole plant, root, leaves	Whole plant is rubbed on the skin to cure fungal infections of the skin. Root powder is good remedy for tooth-ache. Juice made from the leaves destroy the worms in ear.	Hameed et al. (2008)
18	*Phragmites karka* Poaceae	Emergent herb	Leaves, whole plant	Poultice of leaves is used to heal broken bones. Decoction of whole plant is used 2 times a day to cure rheumatic pains.	Sim et al. (2008)
19	*Pistia stratiotes* Araceae	Floating herb	Leaves	Crushed leaves use to cure the Anti-arthritis, anti-inflammatory, curative of Skin disease.	Ikram et al. (2014)

(continued)

Table 6.1 (continued)

Sr. No	Botanical name/ family	Habit	Plant part used	Medicinal uses	References
20	*Plantago major* Plantaginaceae	Emergent herb	Leaves, seeds	Poultice of hot leaves is bound onto cuts and wounds to draw out thorns, splinters and inflammation. Seed powder is taken daily to get rid of intestinal worms.	Najafian et al. (2018)
21	*Potentilla reptans* Rosaceae	Emergent herb	Fruit	Paste made from the fruit of this plant is good in treatment of bleeding wounds.	Launert (1981)
22	*Prunella vulgaris* Lamiaceae	Emergent herb	Leaves	Paste made from fresh leaves is effective remedy for wound healing. Juice gargle 2 times a day to reduce mouth swelling also cures sore throats. Tea made from dried leaves is taken to reduce depression and anxiety.	Meuninck (2016)
23	*Ranunculus repens* Ranunculaceae	Emergent herb	Leaves	A poultice of the chewed leaves has been used in the treatment of sores, muscular aches and rheumatic pains.	Hussain et al. (2011)
24	*Salix babylonica* Salicaceae	Emergent tree	Bark, leaves	The bark decoction is used to relief from pain. Leave decoction is used for lowering fever and for healing infections. Bark has antiseptic properties and is used to clean teeth. Bark is put in the water and that water is used to bathe the newly born babies to avoid any infection.	Abdul qadr (2008) and Kumar et al. (2009)
25	*Schoenoplectus lacustris* Cyperaceae	Emergent herb	Roots	Roots of this plant is converted into decoction on boiling with water and taken to cure constipation and also helpful in urine blockage issues. Also this decoction is given to cancer patient for curing	Ajaib et al. (2014)
26	*Spergularia marina* Caryophyllaceae	Emergent herb	Whole plant	It is commonly used as an antidiabetic.	Sardar (2008) and Sardar et al. (2015)
27	*Stellaria media* Caryophyllaceae	Emergent herb	Whole plant	A decoction of the whole plant is taken internally as a postpartum depurative. Infusion of the whole plant is used in cough.	

(continued)

Table 6.1 (continued)

Sr. No	Botanical name/ family	Habit	Plant part used	Medicinal uses	References
28	*Trapa bispinosa* Trapaceae	Floating herb	Fruits	Fruits are edible, sweet, nutritive, appetizer and diuretic with cooling effect, mainly used to treat Leucorrhoea, seminal weakness, leprosy, fever and diarrhea.	Sardar (2008) and Rahman et al. (2000)
29	*Typha angustifolia* Typhaceae	Emergent herb	Pollen, root	Juice made by pollen is used for the treatment of kidney disorders. An infusion of the root has been used in the treatment of kidney stones.	Kumar et al. (2013)
30	*Veronica beccabunga* Plantaginaceae	Emergent herb	Stems, leaves	Juice extracted from the stems of this plant is taken 2 times a day for diuretic patients. Leaves of the plant are used as poultice on wounds for quick healing. Infusions of the leaves are used to clean the blood. Dried plant is used to make herbal tea which helps in lowering blood pressure and also relief pain during urination	Hughes (1990)

References

Abdulqadr KDS (2008) Pharmacological and antimicrobial studies on aqueous extract of Salix babylonica. Zanco J Med Sci 12:1–11

Ahmed F, Rahman S, Ahmed N, Hossain M, Biswas A, Sarkar S, Banna H, Khatun A (2011) Evaluation of Neolamarckia cadamba (Roxb.) Bosser leaf extract on glucose tolerance in glucose-induced hyperglycemic mice. Afr J Tradit Complement Altern Med 8(1):79–81

Ajaib M, Haider SK, Zikrea A (2014) Ethnobotanical studies of shrubs and trees of Agra valley Parachinar, upper Kurram agency, Pakistan. FUUAST J Biol 4(1):73–81

Akhtar N, Ali SS, Ahmed S, Samin J, Khan MA, Khan MS (2017) Phytochemical analysis oF lepidium didymum. Pak J Weed Sci Res 23(2):155–164

Al-Snafi AE (2015) Therapeutic properties of medicinal plants: a review of their antibacterial activity (part 1). Int J Pharmacol Toxicol 6(3):137–158

Andreasen C, Stryhn H (2008) Increasing weed flora in Danish arable fields and its importance for biodiversity. Weed Res 48(1):1–9

Ansari JA, Inamdar NN (2010) The promise of traditional medicines. IJP Int J Pharmacol 6(6):808–812

Applequist WL, Moerman DE (2011) Yarrow (Achillea millefolium L.): a neglected panacea? A review of ethnobotany, bioactivity, and biomedical research. Econ Bot 65(2):209

Asim ZI, Haq F, Iqbal A (2016) Phenology, life form and leaf spectra of the vegetation of kokarai valley, district swat. J Biol Environ Sci 9:23–31

Ballabh B, Chaurasia OP, Pande PC (2017) Ethnomedicinal plants of Western and Central Himalayas. In: Ethnobotany of India, vol 4. Apple Academic Press, Waretown, pp 149–291

Baytop T (1999) Therapy with medicinal plants in Turkey (past and present). Publication of the Istanbul University, p 312

Bojnanský V, Fargašová A (2007) Atlas of seeds and fruits of Central and East-European flora: the Carpathian Mountains region. Springer Science & Business Media, Dordrecht

Brunel S (2009) Pathway analysis: aquatic plants imported in 10 EPPO countries. EPPO Bull 39(2):201–213

Calabozo B, Barber D, Polo F (2001) Purification and characterization of the main allergen of Plantago lanceolata pollen, Pla l 1. Clin Exp Allergy 31(2):322–330

Capers RS, Selsky R, Bugbee GJ (2010) The relative importance of local conditions and regional processes in structuring aquatic plant communities. Freshw Biol 55(5):952–966

Cassileth BR, Deng G (2004) Complementary and alternative therapies for cancer. Oncologist 9(1):80–89

Chaachouay N, Benkhnigue O, Fadli M, El Ibaoui H, El Ayadi R, Zidane L (2019) Ethnobotanical and ethnopharmacological study of medicinal and aromatic plants used in the treatment of respiratory system disorders in the Moroccan Rif. Ethnobot Res Appl 18:1–16

Chandra S, Rawat DS (2015) Medicinal plants of the family Caryophyllaceae: a review of ethno-medicinal uses and pharmacological properties. Integr Med Res 4(3):123–131

Chen HH, Chiang W, Chang JY, Chien YL, Lee CK, Liu KJ, Kuo CC (2011) Antimutagenic constituents of adlay (Coix lachryma-jobi L. var. ma-yuen Stapf) with potential cancer chemopreventive activity. J Agric Food Chem 59(12):6444–6452

Chen KW, Blatt K, Thomas WR, Swoboda I, Valent P, Valenta R, Vrtala S (2012) Hypoallergenic Der p 1/Der p 2 combination vaccines for immunotherapy of house dust mite allergy. J Allergy Clin Immunol 130(2):435–443

Chiang LC, Cheng HY, Chen CC, Lin CC (2004) In vitro anti-leukemic and antiviral activities of traditionally used medicinal plants in Taiwan. Am J Chin Med 32(05):695–704

Chung CP, Hsia SM, Lee MY, Chen HJ, Cheng F, Chan LC, Chiang W (2011) Gastroprotective activities of adlay (Coix lachryma-jobi L. var. ma-yuen Stapf) on the growth of the stomach cancer AGS cell line and indomethacin-induced gastric ulcers. J Agric Food Chem 59(11):6025–6033

Crosson H (2010) Keeping aquatic plants in their place: common sense tips to protect lakes and rivers. Landscape Online

Dad JM, Khan AB (2013) Assessment of conservation status of five threatened Medicinal Plants at high altitude grassland in Bandipora Kashmir. Med Plants Int J Phytomed Related Industries 5(1):47–51

Dasgupta A, Hammett-Stabler CA (eds) (2011) Herbal supplements: efficacy, toxicity, interactions with western drugs, and effects on clinical laboratory tests. Wiley, New York, pp 202–205

Di Giorgio C, Delmas F, Tueni M, Cheble E, Khalil T, Balansard G (2008) Alternative and complementary antileishmanial treatments: assessment of the antileishmanial activity of 27 Lebanese plants, including 11 endemic species. J Altern Complement Med 14(2):157–162

Dodds W (2002) Freshwater ecology: concepts and environmental applications. Elsevier, Amsterdam

Duke JA (2002) Handbook of medicinal herbs. CRC Press, Boca Raton

Duke JA, Ayensu ES (1985) Medicinal plants of China, vol 2. Reference Publications

Flint NA, Madsen JD (1995) The effect of temperature and daylength on the germination of Potamogeton nodosus tubers. J Freshw Ecol 10(2):125–128

Gan RY, Kuang L, Xu XR, Zhang Y, Xia EQ, Song FL, Li HB (2010) Screening of natural antioxidants from traditional Chinese medicinal plants associated with treatment of rheumatic disease. Molecules 15(9):5988–5997

Ganie AH, Tali BA, Nawchoo IA, Khuroo AA, Reshi ZA, Dar GH (2020) Diversity in medicinal and aromatic flora of the Kashmir Himalaya. In: Biodiversity of the Himalaya: Jammu and Kashmir State. Springer, Singapore, pp 545–563

Ghani KAA, Advia YK (2011) Idara Kitab, al-Shifa. New Delhi, pp 1260–1261
Guarrera PM, Savo V (2016) Wild food plants used in traditional vegetable mixtures in Italy. J Ethnopharmacol 185:202–234
Guil-Guerrero JL (2001) Nutritional composition of Plantago species (P. major L., P. lanceolata L., and P. media L.). Ecol Food Nutr 40(5):481–495
Haider S, Nazreen S, Alam MM, Gupta A, Hamid H, Alam MS (2011) Anti-inflammatory and anti-nociceptive activities of ethanolic extract and its various fractions from Adiantum capillus veneris Linn. J Ethnopharmacol 138(3):741–747
Hamayun M, Khan SA, Sohn EY, Lee IJ (2006) Folk medicinal knowledge and conservation status of some economically valued medicinal plants of District Swat, Pakistan. Lyonia 11(2):101–113
Hameed I, Dastagir G, Hussain F (2008) Nutritional and elemental analyses of some selected medicinal plants of the family Polygonaceae. Pak J Bot 40(6):2493–2502
Hughes RE (1990) The rise and fall of the "antiscorbutics": some notes on the traditional cures for "land scurvy". Med Hist 34(1):52–64
Hussain I, Khattak MUR, Muhammad Z, Khan N, Khan FA, Ullah Z, Haider S (2011) Phytochemicals screening and antimicrobial activities of selected medicinal plants of Khyberpakhtunkhwa Pakistan. Afr J Pharm Pharmacol 5(6):746–750
Ibn Sina AAH (2005) Al-Qanun fi'l-tibb (Canon of medicine). Alamy Le-Al-Matbooat Institute, Beirut, pp 370–385
Ibn Sina A (2007) Al Qanoon Fit Tib (Urdu translation by Ghulam Hussain Kantoori). Idara Kitabul Shifa, New Delhi, pp 212–229
Ikram S, Bhatti KH, Parvaiz M (2014) Ethnobotanical studies of aquatic plants of district Sialkot, Punjab (Pakistan). J Med Plants 2(1):58–63
Isola OI (2013) The "relevance" of the African traditional medicine (alternative medicine) to health care delivery system in Nigeria. J Dev Areas 47:319–338
Jain A, Sundriyal M, Roshnibala S, Kotoky R, Kanjilal PB, Singh HB, Sundriyal RC (2011) Dietary use and conservation concern of edible wetland plants at indo-burma hotspot: a case study from northeast India. J Ethnobiol Ethnomed 7(1):29
Joshi AR, Joshi K (2006) Piscicidal plants of nepal: checklist, ethnobotanical uses and indigenous practices. Ethnobot Leaflets 2006(1):37
Joshi K (2016) Cynoglossum L.: a review on phytochemistry and chemotherapeutic potential. J Pharmacogn Phytochem 5(4):32
Kaigongi M, Musila F (2015) Ethnobotanical study of medicinal plants used by Tharaka people of Kenya. Int J Ethnobiol Ethnomed 1(1):1–8
Kar A (2003) Pharmacognosy and pharmacobiotechnology. New Age International, New Delhi
Keddy PA (2010) Wetland ecology: principles and conservation. Cambridge University Press, Cambridge
Keya MA, Rahman AHMM (2017) Angiosperm diversity at the village Sabgram of Bogra, Bangladesh with emphasis on medicinal plants. Am J Plant Biol 2(1):25–34
Khodaie L, Esnaashari S, Moghaddam SB (2015) Essential oil of arial parts of Adiantum capillus-veneris: chemical composition and antioxidant activity. Jundishapur J Nat Pharm Prod 10(4):3–7
Kırım B, Çoban D, Güler M (2014) Floating aquatic plants and their impact on wetlands in Turkey. In: 2nd international conference-water resources and wetlands. Tulcea, Romania
Kumar KS, Kumar DS, Teja VA, Venkateswarlu V, Kumar MS, Nadendla RR (2013) A review on Typha angustata. Int J Phytopharm 4:277–281
Kumar M, Paul Y, Anand VK (2009) An ethnobotanical study of medicinal plants used by the locals in Kishtwar, Jammu and Kashmir, India. Ethnobot Leaflets 2009(10):5
Lassak EV, McCarthy T (2011) Australian medicinal plants: a complete guide to identification and usage. New Holland Publishers (Australia), Chatswood
Launert E (1981) The Hamlyn guide to edible and medicinal plants. Hamlyn, London
Li L (2000) Opportunity and challenge of traditional Chinese medicine in face of the entrance to WTO (World Trade Organization). Chin J Inf Trad Chin Med 7:7–8

Liakh V, Konechna R, Mylyanych A, Zhurakhivska L, Hubytska I, Novikov V (2020) Caltha palustris. Analytical overview. ScienceRise Pharm Sci 2(24):51–56
Loganayaki N, Siddhuraju P, Manian S (2012) Antioxidant, anti-inflammatory and anti-nociceptive effects of Ammannia baccifera L.(Lythracceae), a folklore medicinal plant. J Ethnopharmacol 140(2):230–233
Ma G, Chong L, Li Z, Cheung AH, Tattersall MH (2009) Anticancer activities of sesquiterpene lactones from Cyathocline purpurea in vitro. Cancer Chemother Pharmacol 64(1):143–152
Mahmood A, Mahmood A, Tabassum A (2011) Ethnomedicinal survey of plants from District Sialkot, Pakistan. J Appl Pharmacol 3:212–220
Malakar C, Choudhury PPN (2015) Pharmacological potentiality and medicinal uses of Ipomoea aquatica Forsk: a review. Asian J Pharm Clin Res 8(2):60–63
Manandhar NP (1986) Ethnobotany of Jumla District, Nepal. Int J Crude Drug Res 24(2):81–89
Manosroi A, Tangjai T, Sutthiwanjampa C, Manosroi W, Werner RG, Götz F, Manosroi J (2016) Hypoglycemic activity and stability enhancement of human insulin–tat mixture loaded in elastic anionic niosomes. Drug Deliv 23(8):3157–3167
Martynov AM, Dul VN, Dargaeva TD, Chuparina EV (2017) Izuchenie himicheskogo sostava travy kaluzhnitsy bolotnoy (Caltha palustris). Voprosy obespecheniia kachestva lekarstvennyh sredstv 4(18):66–71
Meuninck J (2016) Medicinal plants of North America: a field guide. Rowman & Littlefield, Guilford
Mikaili P, Mojaverrostami S, Moloudizargari M, Aghajanshakeri S (2013) Pharmacological and therapeutic effects of Mentha Longifolia L. and its main constituent, menthol. Anc Sci Life 33(2):131
Mollik AH, Hassan AI, Paul TK, Sintaha M, Khaleque HN, Noor FA, Nahar A, Seraj S, Jahan R, Chowdhury MH, Rahmatullah M (2010a) A survey of medicinal plant usage by folk medicinal practitioners in two villages by the Rupsha River in Bagerhat district, Bangladesh. Am Eurasian J Sustain Agric 4:349–357
Mollik MAH, Hossan MS, Paul AK, Taufiq-Ur-Rahman M, Jahan R, Rahmatullah M (2010b) A comparative analysis of medicinal plants used by folk medicinal healers in three districts of Bangladesh and inquiry as to mode of selection of medicinal plants. Ethnobot Res Appl 8:195–218
Mubashir S, Dar MY, Lone BA, Zargar MI, Shah WA (2014) Anthelmintic, antimicrobial, antioxidant and cytotoxic activity of Caltha palustris var. alba Kashmir, India. Chin J Nat Med 12(8):567–572
Munasinghe JU, Dilhan MAAB, Sundarabarathy TV (2010) Utilization of aquatic plants: a method to enhance the productivity of water in seasonal tanks in the Anuradhapura District. In: Proceedings of the national conference on water, food security, and climate change in Sri Lanka, BMICH, Colombo, June 9–11, 2009. Volume 1. Irrigation for food security (Vol. 1, p. 23). IWMI
Najafian Y, Hamedi SS, Farshchi MK, Feyzabadi Z (2018) Plantago major in traditional Persian medicine and modern phytotherapy: a narrative review. Electron Physician 10(2):6390
Narasimhulu G, Mohamed J (2014) Medicinal phytochemical and pharmacological properties of kesum (Polygonum minus Linn.): a mini review. Int J Pharm Pharm Sci 6(4):682–688
Oyedeji AA, Abowei JFN (2012) The classification, distribution, control and economic importance of aquatic plants. Int J Fish Aquatic Sci 1(2):118–128
Pakia M, Cooke JA, Van Staden J (2003) The ethnobotany of the Midzichenda tribes of the coastal forest areas in Kenya: 2. Medicinal plant uses. S Afr J Bot 69(3):382–395
Parrotta JA (2001) Healing plants of peninsular India. CABI Publication, New York, p 143
Peregoy JA, Clarke TC, Jones LI, Stussman BJ, Nahin RL (2014) Regional variation in use of complementary health approaches by US adults. NCHS Data Brief 146:1–8
Petroeschevsky A, Champion PD (2008) Preventing further introduction and spread of aquatic weeds through the ornamental plant trade. In: Brisbane: 16th Australian Weeds Conference, Queensland Weeds Society, pp 399–402

Qureshi RA, Ghufran MA, Gilani SA, Yousaf Z, Abbas G, Batool A (2009) Indigenous medicinal plants used by local women in southern Himalayan regions of Pakistan. Pak J Bot 41(1):19–25

Rahman MM, Mosaddik MA, Wahed MII, Haque ME (2000) Antimicrobial activity and cytotoxicity of Trapa bispinosa. Fitoterapia 71(6):704–706

Ramli MR, Milow P, Chooi OH (2015) Traditional knowledge of a practitioner in medicinal plants of Masjid Ijok village, Perak, Malaysia. Stud Ethno-Med 9(1):59–66

Rasool Hassan BA (2012) Medicinal plants (importance and uses). Pharmaceut Anal Acta 3(10):2153–2435

Riley DS, Riley, Kraplow (2012) Materia medica of new and old homeopathic medicines. Springer, Berlin

Roy DC, Barman SK, Shaik MM (2013) Current updates on Centella asiatica: phytochemistry, pharmacology and traditional uses. Med Plant Res 3

Sampson W (1995) Antiscience trends in the rise of the alternative medicine movement. Ann N Y Acad Sci 775(1):188–197

Santosh K, Satya N (2010) Herbal remedies of wetlands macrophytes in India. Int J Pharm Bio Sci 1(2)

Sardar AA (2008) A contribution to the morphological and ethnopharmacological studies of the pollen grains of wetland plants of Punjab, Pakistan. Doctoral dissertation, GC University Lahore, Pakistan

Sardar AA, Khan Z, Perveen A, Zereen A (2015) Appraisal of ethnobotanical uses of the wetland plants of Punjab, Pakistan. Afr J Tradit Complement Altern Med 12(4):9–13

Sathyavathi R, Janardhan KJ (2011) Folklore medicinal practices of Badaga community in Nilgiri biosphere reserve, Tamilnadu, India. Int J Pharma Res Dev 3(2):50–63

Seo WG, Pae HO, Chai KY, Yun YG, Kwon TH, Chung HT (2000) Inhibitory effects of methanol extract of seeds of job's tears (Coix Lachryma-Jobi L Var. Ma-Yuen) on nitric oxide and superoxide production in raw 264.7 macrophages. Immunopharmacol Immunotoxicol 22(3):545–554

Shaheen S, Bibi Y, Hussain M, Iqbal M, Saira H, Safdar I, Mehboob H, Ain QT, Naseem K, Laraib S (2017) A review on Geranium wallichianum D-Don ex-sweet: an endangered medicinal herb from Himalaya Region. Med Aromat Plants (Los Angles) 6(288):1000288

Shankar K, Liao LP (2004) Traditional systems of medicine. Phys Med Rehabil Clin N Am 15(4):725

Sharma A, del Carmen Flores-Vallejo R, Cardoso-Taketa A, Villarreal ML (2017) Antibacterial activities of medicinal plants used in Mexican traditional medicine. J Ethnopharmacol 208:264–329

Shirazi MAK (1992) Makhzan ol Advieh (The storehouse of medicaments). Tehran University of Medical Sciences, Sahba Press, Tehran

Sim CH, Yusoff MK, Shutes B, Ho SC, Mansor M (2008) Nutrient removal in a pilot and full scale constructed wetland, Putrajaya city, Malaysia. J Environ Manag 88(2):307–317

Sumitra S, Rupinder K, Sharma SK (2012) Pharmacognostical evaluation of Oenothera rosea L'Hér. ex Aiton. root. J Pharm Res 5(6):3269–3271

Swapna MM, Prakashkumar R, Anoop KP, Manju CN, Rajith NP (2011) A review on the medicinal and edible aspects of aquatic and wetland plants of India. J Med Plants Res 5(33):7163–7176

Thong WH, Madhialagan K (2014) Effect of cytokinins on in vitro regeneration of Cardamine hirsuta from nodal explants. Vascular 10:11

Tonny TS, Sultana S, Siddika F (2017) Study on medicinal uses of Persicaria and Rumex species of polygonaceae family. J Pharmacogn Phytochem 6(6):587–600

Tynsong H, Tiwari BK, Lynser MB (2006) Medicinal plants of Meghalaya, India. Medplant Netw News 6(2):7–10

Uddin K, Rahman AHMM, Islam AKMR (2014) Taxonomy and traditional medicine practices of Polygonaceae (smartweed) family at Rajshahi, Bangladesh. Int J Adv Res 2(11):459–469

Vardhana R (2008) Direct uses of medicinal plants and their identification. Sarup & Sons, New Delhi

Warren DM, Green EC (1982) The Techiman-Bono ethnomedical system. In: African health and healing systems: proceedings of a symposium. Crossroads Press, Los Angeles, pp 85–106
World Health Organization (2000) General guidelines for methodologies on research and evaluation of traditional medicine (No. WHO/EDM/TRM/2000.1). World Health Organization
World Health Organization (2001) Legal status of traditional medicine and complementary (No. WHO/EDM/TRM/2001.2). World Health Organization
Xie ZW (1996) Quanguo Zhongcaoyao Huibian (A collection of Chinese herbal drugs). People's Hygenic Publishing House, Beijing, pp 514–515
Xue H, Chen KX, Zhang LQ, Li YM (2019) Review of the ethnopharmacology, phytochemistry, and pharmacology of the genus veronica. Am J Chin Med 47(06):1193–1221
Zabihullah Q, Rashid A, Akhtar N (2006) Ethnobotanical survey in kot Manzaray Baba valley Malakand agency, Pakistan. Pak J Plant Sci 12(2):115–121
Zeller T, Muenstedt K, Stoll C, Schweder J, Senf B, Ruckhaeberle E, Becker S, Serve H, Huebner J (2013) Potential interactions of complementary and alternative medicine with cancer therapy in outpatients with gynecological cancer in a comprehensive cancer center. J Cancer Res Clin Oncol 139(3):357–365

Chapter 7
Threats to Biodiversity of Wetland

Due to human activities many habitats of birds have been destroyed and also through some through natural processes. By studying different reports and articles on wetland it has been showed that throughout the world about 50% zones of wetland have been vanished and still their destruction continues in different nations. Due to human intervention about 95 km^2/year-wetland areas have been lost from 1980s to the beginning of 2000s. Now a day in almost all developed and developing countries wetlands are critically threatened in various ways. The factors that involves the destruction of world's wetlands are: Construction of roads and infrastructure on wetland areas, Agricultural expansion, Cutting of Forests, Industrial pollution, overgrazing, River canalization, over collection of flora and fauna and introduction of exotic species to natural habitat that initiated unevenness in natural wetland zones. The mentioned factors are the main reason for threat as they decrease the number of essential wetland regions of the world, resulting in loss of habitat and degradation (Dudgeon et al. 2006). Due to habitat loss of wetland areas it directly causes adverse effect on birds that fully dependent on wetlands for their existence. Many migrant birds have declined throughout the world due to degradation of some important higher altitude wetlands. Other main threat to wetland is huge amount of industrial waste and pollution reduces natural habitats. A part from water and air pollution some natural disasters and climatic change also pose threats to wetland biodiversity.

(i) **Overexploitation of species** – the loss of species abundance of interacting species and genetic diversity loss is due to the overexploitation of the species. Due to the disturbance in age sex and structural composition the size of the species population get reduced. Over-fishing bases the transformation in the genetic structure of fish populations as the loss of alleles takes place. Hence the genetic diversity gets reduced.

(ii) **Habitat modification** – the modifications occurring in habitat leads to species extinction. This is caused by human activities like construction of dams,

M. A. Butt et al., *Wetland Plants*, https://doi.org/10.1007/978-3-030-69258-2_7

diversion of water for irrigation purposes, cutting of forests and conversion of wetlands or marshy lands for other purposes. Migration of fishes is due to the construction of dams on rivers and shifts populations from their usual spawning grounds and splits the population in two smaller parts. Soil erosion is caused by the excessive cutting of forests resulting into siltation and sedimentation. Due to this it directly affects the breeding process of some animals and also causes gill congestion of small fishes also which results in the death of fishes (Fig. 7.1).

Fig. 7.1 Death of water animals due to water pollution

(iii) **Pollution load** – Four types of pollution can be well-known while studying water pollution:-

(a) **Poisonous pollutants** – Agrochemicals, heavy metals, phenol and acid cause death, if existing in a high amount and affect the reproductive system of several of fish.
(b) **Suspended solids** – this type of pollution directly affects the respiratory system of the fish diversity and also flora of that specific region is also affected by them.
(c) **Sewage and organic toxins** – they cause de-oxygenation due to eutrophication causing death of fishes (Fig. 7.2).
(d) **Thermal pollution** – due to the increase in temperature and decrease in dissolved oxygen concentration the death of some delicate species occur.

(iv) **Lack of Awareness** – in Pakistan, people of all levels are unaware about the economic value and significance of wetlands. So they don't realize the impact of development of wetlands. Due to this unawareness they destroy the biodiversity and wetlands.

All of these factors affect directly or indirectly to the wetland biodiversity. Extreme death of organisms due to these factors may lead to two types of effects

- Extinction of the species
- Reduction in population size of species.

Fig. 7.2 Sewage and organic toxin in water bodies

7.1 Causes of Wetland Degradation in Himalayas

The main reason of wetland degradation is mentioned below:

Destruction of wetland ecosystem can directly damage their functions and the main reason of this destruction is human activities (Fig. 7.3). Construction of roads and diversion of water flow can totally change the hydrological system of soil along with its moisture content (Fig. 7.4). In Pakistan every year natural disasters like floods occurs that is also a serious cause of wetland destruction as it removes soil, loss vegetation that binds the soil and run off pollutants. Over grazing by animals also cause the loss of vegetation in wetlands that result in soil erosion. It has also

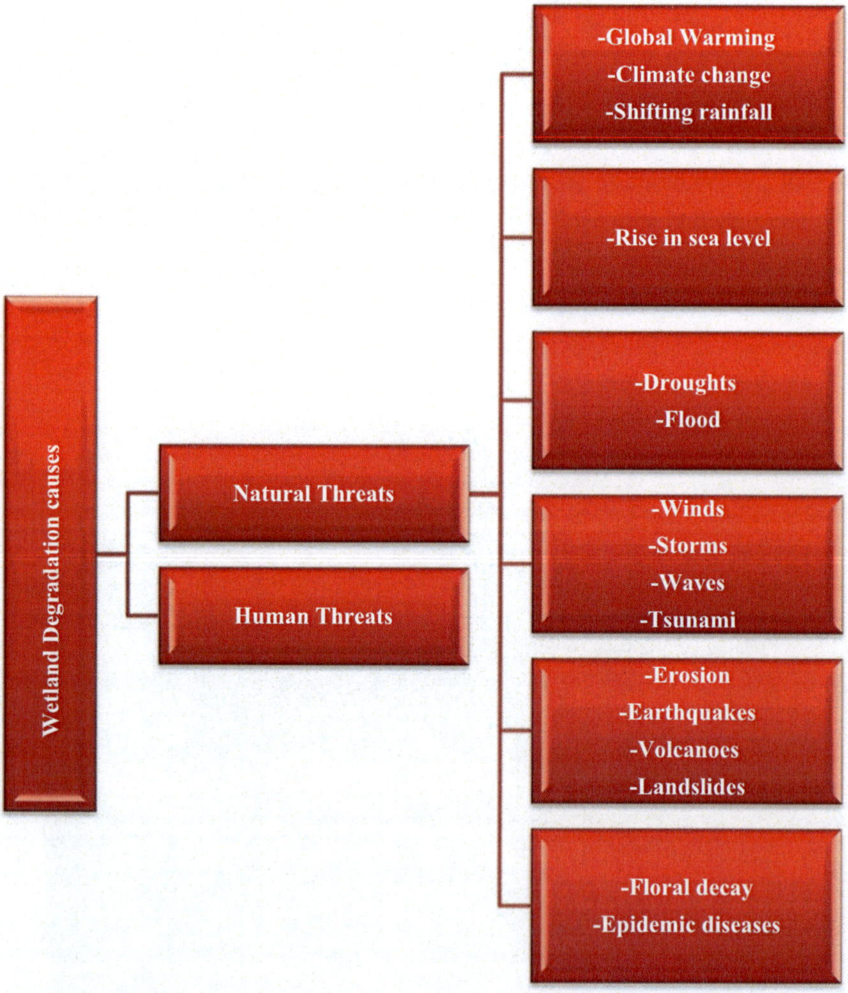

Fig. 7.3 Natural threats causes for wetland degradation

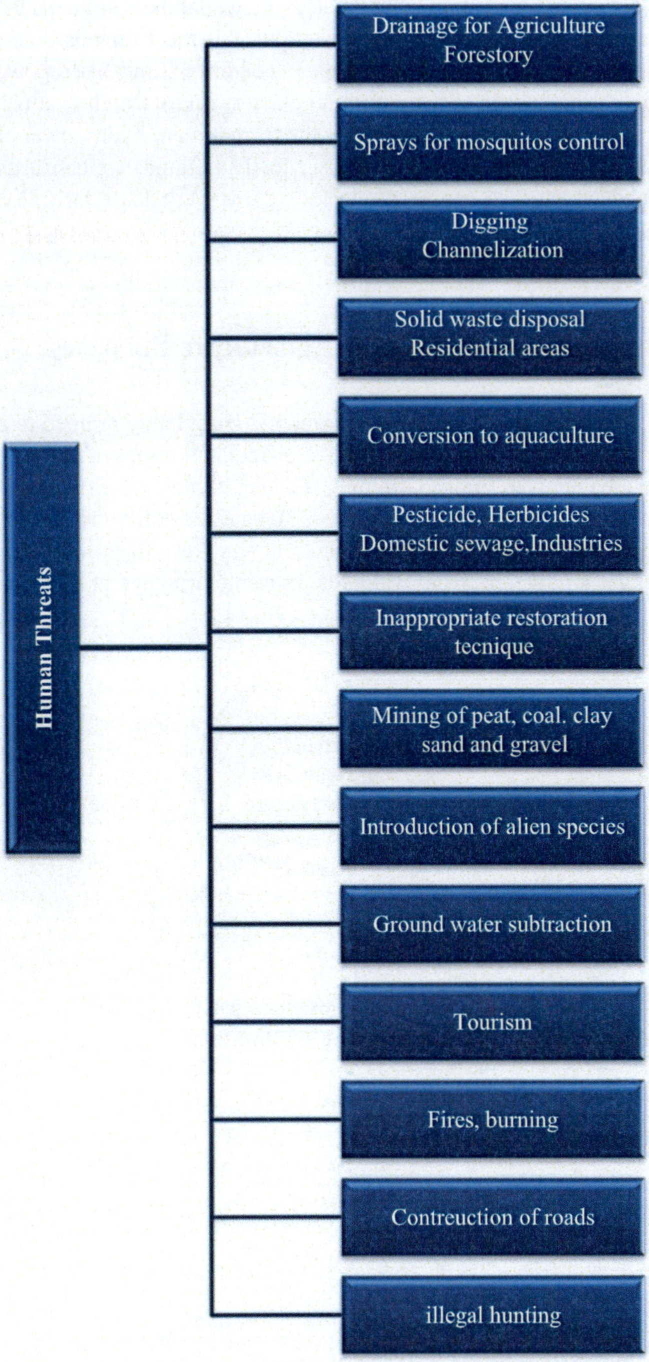

Fig. 7.4 Wetland degradation due to human activities

been got notice that about 182,118 hectors of wetlands is influenced by aquatic weed invasions and also affect the fish output. All the wetlands face an issue of water availability which is due to climatic conditions and hydropower developments. Due to increase in temperature the glaciers melts which results in flooding. Another reason of wetland degradation is the formation of hydropower plants which convert major rivers to seasonal wetlands. Mostly wetlands are converted into agricultural land which also causes the loss of these wetlands (Darwall et al. 2009) (Fig. 7.5).

7.2 Drainage of Wetlands for Agriculture Purposes

Wetlands are under great threat as their water is completely drained to make space for construction or agriculture. Wetlands are frequently viewed as wastelands, and no second idea is given prior to draining the land and clearing its flora to serve the people's requirement and greed (Ayyad 2003). Facilities for tourists are oftenly constructed by draining wetlands. Such human actions ruin the whole ecosystem and we cannot get any benefit from wetlands. Extreme drainage at the corners of peat

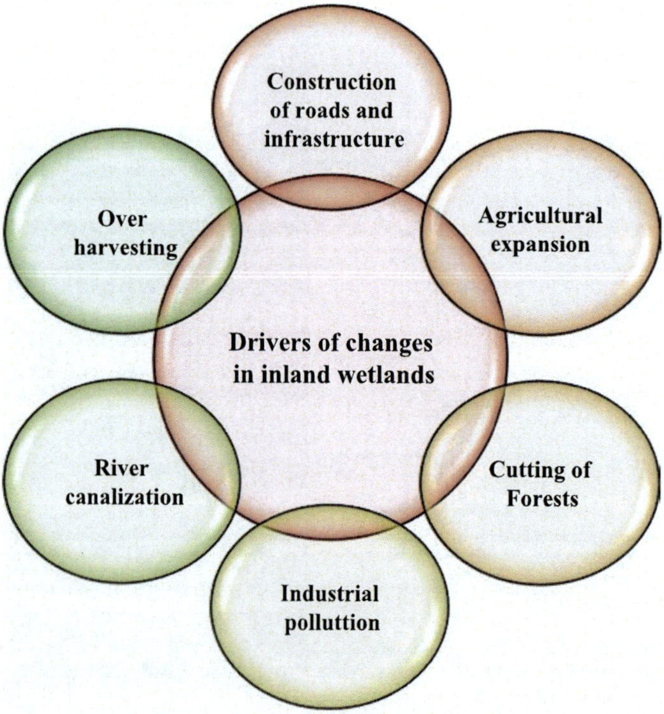

Fig. 7.5 Causes of wetland degradation

bogs is the basic reason for a drop in the water level of wetland's which results in decrease in the formation of peat and changes the varieties of plants that grow there.

7.3 Introduction of Invasive Species

In every ecosystem, the presence of species in wetlands also depends on equilibrium of nature. But due to the introduction of invasive species by humans the whole ecosystem invades, it might lead to a misbalance in the whole system. Competition occurs for natural resources between invasive and native species resulting in destruction of whole ecosystem because of this competition dying of native species takes place (Brinson and Malvárez 2002).

7.4 Water Pollution

Wetlands are usually considered as natural water filtration units. The contaminated water that enters wetlands gets filtered after leaving wetland. During purification water pollutants either settle down in the bottom of wetlands or vegetation of wetlands absorbs this pollutants as nutrients. Though, when the concentration of pollutants go the saturation level, and extra and more poisonous contaminants input the wetlands through water, the species inside the area war to exist. Frequently, wetlands is consider as the waste disposal grounds where industrial wastes, household sewage or wastes is released. Due to huge amount of water pollution resulting in destruction of fauna and flora habitats.

7.5 Climate Change

Due to drastic change in environmental conditions like other habitats, wetlands are also suffering this alteration. Due to increase in temperature the water table of world also increases as a result wetlands become drowned or submerged. Also in some places high temperature results in drought condition and wetlands become completely dry. Therefore, wetlands are getting misplaced both because of higher and decrease water levels triggered by means of the varying consequences of climate variation throughout the world (Tse-ring et al. 2010).

7.6 Construction of Dams

Dam construction has a great impact on the surrounding place specially wetlands where it is built. Dams change the normal flow of a river water to fulfil human requirements. Though, such modifications in the river drift create notable disturbances inside the ecosystems depending on the river. Wetland areas like marshes, swamps, etc., due to this suffers from higher or lower flow of water than normal due to the construction of dams Thus, dams have the energy to permanently harm the wetland ecosystems (Fig. 7.6).

Many pressures have been faced by wetlands only due to human activites, some has indirect adverse effects as sediments and nutrients can swab down from land into the wetland. Pests of weeds and animals effect wetlands. To reduce the effects on wetland careful management should be needed.

7.7 Harvesting and over Hunting

Whitebait, bird hunting and eeling harvesting are common traditions for many people across the world. Harvesting and hunting of these species puts stress on their populations, and also inspires people for wetland habitat conservation. Protecting wetland habitats for harvesting and hunting has spin-offs for many native plants and animals.

In some areas of the world's wetlands peats are harvested for agricultural use. Drains are due to decrease in the level of water and dry out the upper layer of the peat. The flora is removed and the dried out peat skimmed off the upper layer. In some countries this process takes place in fragments, so that some natural flora is left to provide wildlife habitat. When the upper surface of peat is removed, the level of water and natural flora are reinstated to a working bog again. It is not simply the drainage of wetlands or adjoining wetlands that can disturb their level of water.

7.8 Over Grazing

Due to uncontrolled accessibility of live stocks to wetland areas can cause folloeing damages:

- Due to excessive urine and dungs increase nutrient levels.
- Soil erosion takes place.
- Aquatic wildlife disturbance.
- Threats to wetland flora.

Fig. 7.6 Construction of dams and bridges causes wetland destruction

- Weed invasion occurs due to involvement of cattles.
- transfer seeds of weed seeds in their coat, hooves or dung
- Become bogged down and surrounded.

Livestock and Wetlands, can be efficiently protected from each other with the help of stock fencing (Fig. 7.7).

Fig. 7.7 Over grazing of wetland vegetation by cattles

7.9 Increasing Nutrients

Artificial fertilizers that are applied to agricultural fields can disturb surrounding wetlands through nutrient enriched leaching, spray drift and run-off. Entrence of these nutrients to wetlands through floods or storms, after eroded soil swabs down into wetlands. While swamps play a significant role in clearly purifying out nutrients and sediments, excess nutrients in wetlands, specially low nutrient bogs, can alternate the delicate balance of flora and fauna within the wetlands. At the surrounding of peat bogs, improved nutrients support the attack of non-native plants species, such as grey willow.

7.10 Plant Pests

Biodiversity of wetlands in the world is also affected by many plants pests. They can grow rapidly and compete against native plants species for light, space and nutrients. The uncommon grass *Glyceria maxima* is a persistent trouble for replanting programmes in some wetlands and calls for ongoing management to stop it from restricting the fresh native plants. Other pests of plant include numerous species of parrot's feather, yellow flag iris and waterlily.

7.11 Animal Pests

Rabbits, Deer, Pigs and Hares have developed in some wetlands and cause serious threat to native animals and plants. Cats, Rats, hedgehogs, mustelids and possums, are also found in wetlands. These animals are responsible for killing native birds particularly during their breeding period when hatching birds are often reluctant to leave the nest. Eggs of young chicks are also destroyed by native animals such as lizards and bats. Fishes are also affected due to over hunting by humans and over feeding by ducks and goose.

7.12 Fires

In some countries of the world fire naturally occurred in wetlands specially in large peat bogs. Due to this many native plants are adapted to this fire. Some plants are runners that can quickly resprout after a fire. Others depend upon fire for their life as they grow in burn holes.

But wetlands now are a quiet short than they used to be, and a single wild fire can burn most or all of a wetland. When an entire wetland is burnt, few animals and plants start living on to recolonise the burnt region. Animals that survive especially will not live for long without the food and shelter supplied by close by vegetation.

The lack of fires (due to important fire control) way that burn hole habitats are uncommon. This has prompted some flora that rely upon burn holes.

7.13 Wetland Loss: Underlying Causes

Following are the underlying causes for wetland loss:

(i) **Information failures –** Lack of information about conservation and protection of wetland areas among people.
(ii) **Market failures –** The failure of markets to reflect true or full costs of things or facilities provided by preserved wetlands.
(iii) **Intervention failures –** Lack of right combined source management strategies and inter- sectoral rules irregularities leads to wetland degradation or loss.

7.14 Process of Wetland Loss in Himalayas

The wetland loss can be separated into two wide types namely chronic and acute losses. Acute loss includes the filling up of moist areas with soil particles whereas the chronic loss includes the slow removal of forest cover caused by sedimentation and soil erosion of the wetlands over many years (Zaz and Romshoo 2012).

(a) **Acute wetland losses**

- **Agricultural conversion:** In south east Asia there is a culture of rice cultivation which badly effect the wetlands as this crop is totally wetland dependent activity and grow in riparian regions, savannah zones and river deltas. Due to captured rain for making fish pond aquaculture in the catchment regions and rice cultivated parts are not wetlands, water is deprived to the downstream natural wetlands.
- **Deforestation in wetlands:** Vegetation grow at the surrounding of water are used for livestock fodder, wood for building materials, fuel wood, medicinal purposes, beeswax honey and for extracting compounds for leather tanning. Due to construction of excessive amount of fish ponds and alternative farming methods thick wetland vegetation has to be removed which cause serious threats to natural wetland ecosystem. Important function of water ecosystem is to protect against storm and floods but the production of fish farms caused increased pollution load on water like increased amount of organic wastes, lime, chemicals, pesticides, disease causing creatures. This can not only effect water but also effects directly dependent on these natural materials, fish proteins and income (Fig. 7.8).
- **Hydrologic alteration:** Change in the hydrology can change the functions, character, appearance and the values of wetlands. The changes in hydrology encompass either the removal of water from wetlands or elevating the and-surface elevation, such that no floods occur. Initial growth within the crop productiveness has given way for decreased salt accumulations and fertility in soil due to wet farming of arid soils.
- **Inundation by dammed reservoirs:** By holding the water, hydrology of a region is changed and allows for connecting moving water as a cause of energy. While the advantages of energy are well familiar, it also changes the ecosystem.

(b) **Chronic Wetland Losses**

- **Changes of upper watersheds:** Watershed situations have an effect on the wetlands. The circumstance of the land where rain falls, assembles and runs-off into the soil will affect the hydrologic regime and the character of the downstream wetlands. When deforestation, overgrazing and agricultural practices eliminates the water-holding ability of the soil then soil erosion becomes more obvious.

Fig. 7.8 Deforestation of wetlands

- **Extinction of native flora:** Wetlands of Himalays support great number of species of animals and plants. However habitat loss have threatened the diversity of these ecosystems (Gopal 2013). Introduction of invasive species like *Eichornia crassipes* and *Salvinia molesta* have threatened the wetlands and blocked the water channels competing with the native flora. In a current, effort for prioritization of wetlands for protection, Samant (1999) mentioned in his study that as many as 700 wetlands do not have any facts to prioritize. Many of those wetlands are disappearing (Fig. 7.9).

Fig. 7.9 Introduction of exotic spp (*Eichornia crassipes*) to wetland

- **Urbanization:** Wetlands nearby cities are under great developmental burden for industrial, commercial and residential services. Wetlands at the surrounding of cities are important for conserving public waters supplies.
- **Anthropogenic activities:** Unplanned agricultural and urban growth industries, impoundment, source removal, dredge disposal and road construction cause wetlands to be transformed and drained, affecting significant ecological and economic losses in the long term.
- **Hydrological activities:** diversion of rivers and stream and construction of canals to irrigate lowerlands for agricultural purposes has changed the drainage pattern thus destroy the whole weland system.
- **Deforestation:** Soil erosion and siltation occur due to the removal of vegetation in the catchments.
- **Pollution:** pollution due to the excess of agricultural activities, leakage from poisoned water, and polluted stormwater can overcome the purifying capability of wetlands; impacting downstream shoreline waters. Continuous removal of sewage water and toxic elements from industries has contaminated many freshwater wetlands (Fig. 7.10).
- **Salinization:** The excessive extraction of groundwater leads to salinization.
- **Aquaculture:** Huge demand for fishes may results in the conversion of several wetlands into fish ponds and fish farms (Fig. 7.11).
- **Non-native predators:** Mongooses, Feral cats, cattle egrets, rodents, and owls eat native ground-nesting birds and their babies. Toads and Bullfrogs eat eggs of fish, insects, and even birds. These predators can destroy the population of native birds. Non-native plants such as grasses and non-native

Fig. 7.10 Water pollution caused by human activities

weeds compete with native plants and grow colonies in wetlands, which results in the reduction of plant diversity in wetland and also affect ecological functions. Introduction of non-native fish that prey on damselflies and consume wetland vegetation, thus reducing food obtainability for water birds.

- **Hybridization:** Some endemic animals are currently in danger of extinction due to cross-breeding.

Fig. 7.11 Aquaculture ponds

- **Climate Change:** Increases in temperature due to global warming results in an increase in the level of water which affects the whole water system. Future fluctuations in local rainfall and increased temperatures will affect bogs vegetation at higher altitudes, and seasonal wetlands result in drying.

References

Ayyad MA (2003) Case studies in the conservation of biodiversity: degradation and threats. J Arid Environ 54(1):165–182

Brinson MM, Malvárez AI (2002) Temperate freshwater wetlands: types, status, and threats. Environ Conserv 29:115–133

Darwall W, Smith K, Allen D, Seddon M, Reid GM, Clausnitzer V, Kalkman VJ (2009) Freshwater biodiversity: a hidden resource under threat. Wildlife in a changing world – an analysis of the 2008 IUCN Red List of Threatened Species, p 43

Dudgeon D, Arthington AH, Gessner MO, Kawabata ZI, Knowler DJ, Lévêque C, Sullivan CA (2006) Freshwater biodiversity: importance, threats, status and conservation challenges. Biol Rev 81(2):163–182

Gopal B (2013) Future of wetlands in tropical and subtropical Asia, especially in the face of climate change. Aquatic Sci 75(1):39–61

Samant S (1999) Prioritization of biological conservation sites in India wetland. Setting biodiversity conservation priorities for India. World Wide Fund for Nature, New Delhi, pp 155–167

Tse-ring K, Sharma E, Chettri N, Shrestha AB (2010) Climate change vulnerability of mountain ecosystems in the Eastern Himalayas. International Centre for Integrated Mountain Development (ICIMOD), Kathmandu

Zaz SN, Romshoo SA (2012) Assessing the geoindicators of land degradation in the Kashmir Himalayan region, India. Nat Hazards 64(2):1219–1245

Chapter 8
Conservation Strategies of Wetland Flora

8.1 Conservation Strategies of Wetland Flora

A sustainable developmental strategy is required which classified wetlands on the basis of their values to conserve their biodiversity. This classification is based on: for each wetland an Environmental Capital Index might be prepared to concentrate on the local, national, and global standards of the wetlands. Wetlands should then be positioned into one of 3 categories consistent with their Index, viz:

(i) **Natural Capital Stock:** The category with the best Index constitutes the Natural Capital Stock of the state and thus, has the very best conservation requirements.
(ii) **Natural Exploitable Stock:** This category will typically contain the biggest proportion of wetlands. The vital capabilities and values of these wetlands might be retained however others, deemed non-essential, can be allowed to degenerate.
(iii) Natural Replaceable Stock. This is the bottom category containing the one's wetlands of minimum cost which may be destroyed. No net loss policy which is pioneered by the USA needs to be adopted to retain the total wetland resource. Thus, if a wetland is destroyed, some other similar area has to be created. The biodiversity and functional values of the newly-created wetland won't be equal but, with time, it's going to improve its own features and standards. Because of the overall cost of wetlands to the global environment, the precautionary argument demands that each one decision and evaluation veer closer to protecting the habitats until expertise may advocate otherwise.

To classify the wetlands it is essential to produce management plans suitable to their requirements. The capability of all sections of wetlands which includes, grazing, fisheries, seasonal agriculture, cropping, wildlife, tourism, floral vegetation, etc. must be evaluated for the sustainable exploitation of wetlands. Targets and aims

M. A. Butt et al., *Wetland Plants*, https://doi.org/10.1007/978-3-030-69258-2_8

are required to enhance the benefits of the wetland for specific functions and this needs to be monitored for the establishment of management plans. If the objectives are not achieved, the plan has to be changed before the severe loss occurs.

Following steps has to be followed for the conservation and wise use of wetlands:

- Create a Wetland List for the country.
- Consider the benefits of the wetland in relation to biodiversity, anthropogenic and functional values.
- Categorize each wetland on the basis of the Environmental Capital Index.
- Wetlands are categorized into the following: Natural Capital Stock, Natural Exploitable Stock, and Natural Replaceable Stock.
- For each category of wetland management plans has to be set to conserve wetlands. To fulfill these plans targeted work has been done.
- Surveys have been conducted to monitor whether the targets have been achieved or not.
- Advanced methods of management have been used to achieve the targets.
- Stand by the principle of sustainable development.

All the points given above when followed properly the strategy of wetlands conservation have been successfully achieved.

8.2 Ramsar – The Convention on Conservation of Wetlands

In 1971 the Convention on wetlands was held in Ramsar which is the city of Iranian shores. This convention is of international importance as it mainly focuses on the waterfowl habitat. The Convention becomes astonishingly far-sighted for its time, spotting several crucial principles which can be now broadly accepted: the interdependence of man and surroundings; the important ecological role of wetlands as managers of water systems; and the worth of wetlands in terms of economics, science, and culture. This situation with the functioning of wetlands, and the way it impacts mankind and his economic and cultural well-being, has turn into more important over the starting 25 years of Ramsar and will absolutely be major trouble for the twenty-first century, while water supply becomes even more crucial.

The major focus of this convention was on the waterfowl habitat, though Ramsar has become an international tool that deals with wetlands. Due to this Ramsar becomes the only international convention that deals with the particular type of ecosystem i.e. wetlands – rather than on other issues or on single species. This method is purely natural and gives a wide idea that forest and wetlands are the most threatened ecosystems of the world.

Establishment of Ramsar international convention is based upon two concepts:

- To prepare the list of internationally important wetlands prepared by members of government who have the right to accept the responsibility of maintaining ecological characters of these areas.

- The second concept is to wisely utilize the wetlands in the territory of a contracting party. The wise use of wetlands is the term used for the sustainable use of wetlands.

Recently the term wise use of wetland become more important, as due to increasing awareness it is noticed that listed wetland sites are affected by decisions that are taken outside the boundaries which result in the crucial requirement of conservation and wise use of wetlands. Whereas the Ramsar text sets out basic concepts, guidance is required on how to put them into practice and how to adapt them to change the whole world's opinions.

To overcome all these difficulties the convention has made measures to guide the members of the state to carry their responsibilities to list and conserve internationally important wetlands. The responsibilities include:

- The Criteria for Identifying Wetlands of International Importance, which identify sites that could be included in the List; actual designation for the Ramsar List remains the prerogative of each Contracting Party concerned. The various categories of criteria, which have been developed over the years and are still evolving, cover: representative or unique wetland ecosystems; wetlands of value because of the diversity of their plant or animal life or because they support threatened or endemic species; and wetlands which are of particular value for waterfowl (high numbers of a range of species or 1% of the total numbers of one species, sub-species or population), or for fish.
- A simple, global Classification System for Wetland Types which identifies 35 types of wetlands, collectively with a wetland datasheet, so that all the world's Ramsar sites may be labeled and defined in a popular fashion. These descriptions are stored within the Ramsar Database, which can be used to analyze widely differing wetlands across the world, and as a basis for the recommendation to wetland managers, drawing on experience at comparable websites in other corners of the world.
- The Montreux Record, which identifies Ramsar sites whose biodiversity, is beneath unique pressure within the face of technological developments, pollutants, or other human interference.
- The Management Guidance Procedure (formerly referred to as the Monitoring Procedure) which permits the Ramsar Bureau to arrange missions to visit Ramsar web sites (mainly those at the Montreux Record) and to offer advice. The Procedure has been operated at more than 30 Ramsar web sites and more than a few brilliant successes in the conservation of wetland biodiversity had been achieved.

8.3 Future Planning of Ramsar

A range of objectives and actions have been identified to assist Ramsar to obtain its assignment in this period and some of the most critical are highlighted below:

- The Convention can only be fully effective if as many states as possible become Contracting parties, so there may be a strong drive to recruit new members for Ramsar, particularly in under-represented areas such as the Near East, Caribbean, the Pacific, and Southern Africa, with an aim of 120 members by 2002.
- Subsequently, National Wetland Policies are important to the purpose of accomplishing wise use; importance will be placed on establishing a larger number of National Wetland Policies.
- Applications of the Convention will attention to the function of wetlands in the background of land-use making plans (in particular in the coastal zone making plans and river basin management). Special attention could be given to the connection between wetlands and water useful resource management. For many developing countries, water could be one of the scarcest and maximum precious resources inside the twenty-first century, yet wetland conservation and water supply, which are very intently associated subjects, are often dealt with totally unrelated sectors.
- More particular ways of calculating the real financial values of wetland capabilities will be provided; greater attention might be paid to wetland recuperation and rehabilitation, to the empowerment of local communities in wetland management, and to the involvement of the private region in wetland issues.
- Another major assignment might be to build on the work of regional wetland inventories and individual wetland scientists, to offer a better typical definition of worldwide wetland sources and consequently the scope of the Convention's work.
- As for Ramsar sites, attention might be paid both to quality and quantity. The goal is to reach 1000 Ramsar web sites by means of the year 2002 and to ensure that they are properly managed (50% need to have control plans by way of then) and that their ecological person is maintained and monitored. In destiny, extra emphasis might be placed on designated sites from sure wetland types hitherto given insufficient attention, appreciably mangroves, peatlands, coral reefs, and seagrass beds.
- Greater public guide for wetland conservation and wise use may be sought by more intensive efforts to raise recognition of wetland values and functions. A natural consequence of this method may be to reinforce the capability of institutions, specifically in growing countries, through training programs, and to ensure those development agencies, each multilateral and bilateral, take account of wetland values, and a part of the mission of raising consciousness.
- Ramsar receives many needs for investment of initiatives on conservation and smart use and an attempt could be made to grow the resources to be had to the Small Grants Fund to 1,000,000 bucks a year. In addition, the Bureau will act as a catalyst, identifying investment assets for projects on conservation and sensible use of wetlands, whether or not with bilateral or multilateral investment agencies, non-public sector sources, foundations, or NGO bodies.

8.4 Ramsar Convention in Asia

Two successive Meetings of the Conference of the Parties to the Ramsar Convention in influential international locations within the location, namely Japan in 1993 and Australia in 1996 have been powerful in raising the notice in many Asian countries of wetland conservation in standard and the Ramsar Convention in particular. To add to this growing awareness, there have been various environmental training campaigns relating to wetland conservation in many countries within the vicinity. A superb end result of that is that the number of Contracting Parties to the Convention is growing: the Governments of Cambodia, Republic of Korea, and Thailand have officially introduced their aim to become members of the Convention and Bhutan has knowledgeable the Ramsar Bureau that it is getting ready to do so. A variety of Contracting Parties, which includes China and Malaysia, as well as non-member states are within the method of developing National Wetland Policies, Strategies, or Action Plans. However, the increase in the range of Ramsar sites within the location is as a substitute slow and governments, NGOs, and the personal zone will need to make superb efforts to preserve the ecological person of precious wetlands in the place.

8.5 Laws for Wetlands

Nature conservation laws have tended to evolve along comparable lines around the world even though with versions of emphasis and to very different ranges of advancement. Their scope and objectives normally involve development through the following stages:

- Rules and regulations for fishing and hunting.
- Formation of protected areas and protection of threatened species.
- More comprehensive protection of certain groups of animals, firstly birds, and mammals (excluding harmful species), then reptiles and amphibians some species of plants and invertebrates.
- Conservation of certain categories of natural habitat.
- Incorporation of preservation of the natural environment inside local development and land-use law.
- Identification, regulation, and control of techniques which adversely have an effect on biological diversity.

8.6 Conservation Strategies Based on Public Acquisition of Wetlands

In many countries, public ownership under the precise control regime is considered the main assurance of making sure about the security and improvement of main wetlands and protecting them from challenging short-term financial interests. A near-equivalent method is to impose very strict planning restrictions on privately owned wetlands to prevent the change of land use or full-size degradation.

Under such methods, importance is given to three groups of natural regions:

- Those that carry out important ecological functions inside the water cycle and provide significant economic benefits, such as the floodplains.
- As a result of intense pressure due to economic development, the growth of population, and many other works that demand land resources could affect the ecological functions. Densely populate or huge urbanized areas along coastal zones, particularly the nearly surrounded Mediterranean sea, is the typical illustration in this regard: Public acquisition may be the most effective way to confer long-term protection and is likelier to be quicker – if costlier – than instituting negotiations for the demarcation and ultimate establishment of a protected area.
- Those located in major zones of a protected area (in order to assure its strict conservation) or round a protected place. In the latter case, public acquisition can be used to grow the total region managed for conservation and consequently secure constant land and water use over a bigger share of the ecological unit or catchment place.

8.7 Legislation for the Conservation of Wetland Habitat Types

A developing number of nations have received or altered laws to give defensive status on certain habitat types; anywhere they take place, generally to check the alarming rate of their loss. Such laws usually restrict the demolition or modification of these habitats apart from grants. The habitat forms secured from this method are not of course restricted to wetlands: many other types of semi-natural or natural ecosystems, such as old-growth forest and grasslands, are under risk and may be secured by this technique. In any case, most current laws in this classification explicitly address the insurance of probably a few sorts of wetlands.

Habitat conserving methods are quite new and still, they are developing day by day. In certain nations, their degree and techniques have been dynamically corrected by means of rules and regulations in order to increase transparency and efficiency.

Following are the list of key factors which help in conservation of wetland habitat type:

- Nature of enabling legislation (past systems were established under nature conservation law, however new systems may be established under framework environmental law or on water laws).
- Institutional competence (in nations with a federal or decentralized structure of government, duty for implementing or making legislation may be divided between subnational and national governments or only given to sub-national level).
- Range of habitat types covered (laws are applied only to marshes and lakes as a minimum but some go much further).
- The minimum size of wetlands benefiting from legal protection.
- Ownership of wetlands subject to the legislation (allow structures related to habitat types can be restrained to privately-owned land or apply for no matter land ownership).
- Functioning effects (Some legal guidelines are sufficiently designated to provide a potential regulatory system, whilst others lay down common rules which must be implemented via secondary site-precise regulations).

8.8 Conservation of Particular Categories of Wetlands

8.8.1 Conservation Laws for Lakes and Ponds

Legally, lakes were not the part of wetland category. Therefore, there are no separate rules and regulations for conservation like other categories of wetland there are rarely any particular rules applicable to lakes other than those for the protection of any class of wetland falling inside the definition given by national legislation. Where lakeshores are secured, they are typically dependent upon, the same rules as for river banks. In developing countries like the United States, laws for conserving inland wetlands are just as applicable to the lakes.

8.8.2 Conservation of Peat Bogs

This type of wetland is the most under-represented on the Ramsar List and is threatened by artificial reforestation, drainage, eutrophication, and the removal of peat: before the ecological significance of peat bog habitats was known the extraction of peat is lawfully carried out with the approval of the government. This results in a significant reduction of water to peat bogs and causes serious damage to them and their ecological functioning. Upon rehydration, the peatlands can hardly be reconstructed. Their compaction, shrinkage, and destruction leads to soil erosion and harms the flood control function.

Peatbogs are similarly simple to delimit and, except if they are found in forests, commonly come under the department of nature conservation. They may be secure in a site-specific way by the formation of nature reserves or in usually applicable rules. According to rules of nature conservation act in Upper Austria, special permission has been taken from the nature conservation authorities for the afforestation and drainage of peat bogs and marshes which are larger than 5 hectares. Also, permission is required for the extract of peats. In Sweden, peat is used for the production of energy before the implementation of the peat act on 13 June 1985, which bans the utilization of peat bogs for energy production without a permit. Permission for energy production may be granted where the suggested plan is in the favor of public interest. The permission is only granted if the damage or destruction caused by this is low and should be benefitting the local population. The worker must submit a bond as a security to guarantee that the conditions will be followed. To fulfill other requirements by utilizing peat permission should be taken from authorities and also ensure that this project has a low adverse effect and of public interest. In Italy, legislation adopted on 3 November 1998 to control peat bogs extraction and to re-construct the damaged habitat types.

8.8.3 Glaciers

Specific protection law for the conservation of glaciers was established by the Austrian government in Carinthia, Vorarlberg, and Tyrol. Strict rules have been implemented by Carinthia's Nature Protection on 3 June 1986 to prohibits any modification or alteration of glaciers and their upper watershed. There are a few opportunities for exclusions however these are restricted. In France, on 9 Jan 1985, the Mountain Development and Protection Act provide territorial land-use plans in mountain zones that may designate protected glaciers. According to Galasso law, any change and modification in the state of glaciers and its circuses should require special permission as it is included in their law which was established on 8 August 1985.

8.8.4 Conservation of Mangroves and Other Coastal Wetlands

Marine and coastal wetlands secured by wide Ramsar definition which includes deltas, estuaries, coastal marshes, mangroves, lagoons, mudflats, and coral reefs. Economically many of these habitat types are of great importance which includes tourism, forest products, and fisheries. Mostly because of jurisdictional imperatives, very little protection has been given to marine or coastal habitats under national and subnational law systems.

However, legislation for the protection of marine and coastal habitats passed in some but growing number of countries. The constitutions of some Brazilian states incorporate mangroves and other seaside wetland types among the list of protected habitats; however, this protection depends on the practical implementation of legislation. These laws are also followed in countries like the USA, Spain, Costa Rica, and France. Before this, another Act of coastal planning and protection was made on 3 January 1986 which particularly covers coastal heaths, sand dunes, mudflats, coastal woodlands, beaches, marshes, and uninhabited islets. In addition to this mangroves and coral reefs are also protected.

In 1985 the act of Canadian fisheries bans, excluding on permission work, or any kind of undertaking which results in harmful change, destruction or disruption of breeding lands and nurseries, food supply zones upon which marine animals directly or indirectly depend to carry their normal life functions. Before the beginning of any activity, the impact of projects must be considered as they are affecting the habitats of fish. Due to this act, any government would take permission for the filling or digging of coastal wetlands: thus the removal or destruction, cutting of mangrove trees is also banned in protected areas except under license.

Mangrove environments are in an exceptional situation concerning the law since they are secured with trees, which are subjected to forestry legislation and rarely take account of the unique character of mangroves as coastal wetlands. Indeed, even where mangroves are considered as forest reserves, this may not be enough to completely reserve mangrove ecosystems or even to secure them against the illegal cutting. Special legislation to preserve mangroves biodiversity has been developed by several countries namely; Philippines, Senegal, and Thailand following forest oriented approaches. In 1972 Venezuela included in the list of those countries which develop separate legislation to conserve mangrove's ecosystem. The Mangrove Protection Decree requires any individual or legitimate element aiming to complete activities, projects, and building works at risk to influence mangrove ecosystems and the related condition to acquire a license from the governmental authorities. An ecological effect appraisal must be conducted and the grant may possibly be allowed if the specialized evaluation fulfills four conditions.

- The Ministry of Environment must confirm that there is no other practicable area for the planned activities.
- The project should include low damage to the ecosystem.
- The normal currents of the river and marine water must not be disturbed.
- The project manager must ensure to minimize and correct any loss to the environment.

Certain activities were banned in mangrove ecosystems which includes;

- The use of pesticides must be prohibited excluding epidemics.
- Construction of floating houses or building of houses on pillars.
- Planting of alien species by replacing mangroves.
- The release of liquid effluents or building refuses into mangroves.
- All those activities which according to the ministry of Environment declares to be destructive for the growth of mangroves or their dependent or associated species.

Exceptions from these restrictions might be made in two cases.

- Firstly, special administrative zones have been designated by the expert authorities for permanent forestry which is excluded from these necessities, given that administration plans have been set up for the region being referred to.
- Secondly, people living closer to protected areas that rely upon ecosystems for their survival, which includes fishermen, providing that their actions do not damage the mangroves.

8.8.5 Flood Plains and River Banks Conservation

As freshwater ecosystems include river banks, riparian flora, and flood plains, combine rules and regulations should logically embrace these factors of biodiversity. Under the legacy of river management in some countries, clearance of bank vegetation is still required. In developed countries of Europe; charges were towed by horses on paths along the bank of the river. This made it important to build up an open option to proceed along navigable waterways, in which diversion paths of water, obstacles, and fences were disallowed and the growth of trees was not permitted. In addition to this, due to thick and dense vegetation along the banks, it stops free-flowing water during flood conditions. Since it is frequently important to dig a riverbed with the purpose of channel formation and for the formation of an obstruction-free strip of land along the banks to allow for its passage heavy equipment may be required.

On the other hand, it has been progressively understood that the support of riverbanks in a characteristic state has numerous environmental favorable circumstances just as contributing to landscape quality. Riverbanks with diverse flora are the habitat for many wetland species, for example, otters; these banks also act as natural corridors for wildlife; they save waterway waters against hazardous pollutants, mainly pesticides, insecticides and artificial fertilizers from adjoining areas, and therefore they act as barriers to protect rivers from adverse outside effects. Thus the freshwater ecosystem plays an important role in restoring and maintaining ecological stability.

To conserve wetland first option is to form a strip along the riversides which protect the vegetation, the length and width of this strip are according to the width of the watercourse, and also it varies from country to country. Conditions are controlled within this protection strip, as the trees and vegetation fall in this strip are not allowed to cut, browsing and grazing may also be prohibited in such areas. This type of law is required for dealing with water or rivers or wet areas alongside. A 2nd choice is to establish progressively broad land-use controls, especially on development, inside security strips along riverbanks. The 3rd option includes the extension of legitimate security to flood fields. Recently, these were broadly seen as territories that only should have been protected, drained, and developed from floods.

8.9 Biodiversity Conservation in Wetlands of Pakistan: An Overview

In Pakistan, much work has been conducted related to the conservation of biodiversity of plants. Chaudhri and Qureshi (1991) reported 709 vascular plant species from Pakistan, which are being threatened. Shirazi (1993) wrote a book based on the conservation of the Wetlands of Pakistan. Martin (2014) conducted research on trees and shrubs of the Suleiman Range which are used by humans to fulfill their fuel needs thus at this region the vegetation is critically threatened. Pimbert et al. (1996) conducted a study on the wetland management policies in Pakistan and the neighboring country India. Khan (1997) reported the endangered wetlands of Pakistan along with their conservation strategies and also studies the conservation of wildlife biodiversity in these areas. Some avian species are also linked with wetlands for food and shelter work has been done on the ecology distribution and conservation of these species (Ali 2005). Baloch and Tanık (2008) worked on the development of water areas with special emphasis on the conservation of wetlands in the Baluchistan province of Pakistan. Khan and Baig (2017) worked on the wetlands of Pakistan, his work mainly focuses on the distribution degradation and the conservation of wetlands.

Chaudhry (2010) reported the threatened areas of wetlands in Pakistan and the reasons where animal and plant biodiversity threatened in these areas. Gill et al. (2012) study on the conservation of avian species linked with the wetlands and also studies the impact of wetland degradation on the breeding of these species in Chakwal, Pakistan. Qureshi and Ali (2011) discusses the effect of climatic change on the biodiversity of wetlands in Pakistan. Gill et al. (2012) reported the effect of saline soil on the wetland floral diversity in the Cholistan Desert. Khan and Baig (2017) study the higher altitudes wetlands of the Hindu Kush Himalayan region of Pakistan and the challenges faced by the biodiversity of these regions.

8.10 Himalayan Wetlands Conservation

The Himalayan wetlands are under threat due to unregulated urbanization and unsustainable tourism. Urgent attention at the policy level is the need of the hour.

Himalayan wetlands continue to receive very little policy attention or investment. The Wetland (Conservation and Management) Rules, 2010 was replaced by the Wetland (Conservation and Management) Rules, 2017. High-altitude wetlands do not feature in this revised policy.

Climate change is predicted to have a major influence on the Himalayan region. The Himalayas are expected to experience higher levels of precipitation and warming that can lead to faster melting of snow and receding of glaciers. These changes will also have an impact on the vegetation of the region. Some signs are already obvious. There has been an increase in floods, cloudbursts that have impacted the wetlands. Rising temperatures, increase in pollution have also led to the destruction of flora and fauna and also impacted the livelihoods of communities depending on the wetlands in the region.

Besides climate change, a number of factors such as land use and land cover changes, urbanization, increase in the tourism industry, the setting of army camps and their interventions in the area, especially in places like Ladakh, are all taking a toll on the wetland ecosystems in the hill states.

A pilot project on conserving high-altitude wetlands and lakes in the Himalayan–Karakorum–Hindu Kush (HKH) region in Pakistan, initiated in 2004, has adopted a participatory approach involving local communities. So far 6 management plans have been developed. Of these, 2 for Handrap and Utter lakes will be implemented in the next 5 years in a partnership approach involving local communities, forest, wildlife, and fisheries departments of the Northern Areas. As part of preparing the management plans and improving the status of the sites, several activities have been undertaken. Efforts include:

- Helping communities conduct their own environmental monitoring surveys in Handrap and Utter lakes and their immediate catchments. Integrating this information in the management plans;
- A joint ecological monitoring survey of flora and fauna conducted at Ishkoman Community Controlled Hunting Area;
- Campaigns to keep the lake areas clean, organized with stakeholders at Handrap and Utter lakes;
- Developing a GIS database of 12 potential high-altitude wetland sites;
- Developing GIS-based land use maps for 12 potential HAWs and their catchments areas;
- Refining and standardizing HAW inventory techniques and sharing with partners for adoption;
- Conducting training and awareness-building activities to conserve the sites by providing vital information to the general public.

WWF Pakistan has prepared plans for 2007–2012 to continue the implementation of management plans for Handrap and Utter lakes and develop participatory HAW conservation models for replication in the rest of 10 potential sites.

8.11 Methods of Conserving Floristic Biodiversity

To conserve the floristic biodiversity following efforts are made:

8.11.1 Ex-situ *Conservation*

Growth and propagation of various species under controlled laboratory conditions is termed as *ex-situ* conservation. It is an effective technique to conserve endangered and threatened plant species outside their natural habitats. It includes the protection of endangered plant species in laboratories, botanic gardens, gene banks, or in plant nurseries. The *ex-situ* conservation method is beneficial when the population of a species is near extinction. In Pakistan, many threatened plants are grown in laboratories such as *Geranium wallichianum*, *Paeonia emodi*, *Sassurea lappa,* etc. as they are medicinally important and some seeds of different endangered trees were also collected and nurseries were built with the collaboration with the forest department. Many institutes and organizations in Pakistan work to conserve biodiversity as Plant Genetic Resource Institute (PGRI) has country largest seed store bank and most water plants or aquatic plants are provided controlled conditions in National Agricultural Research Center (NARC), Islamabad as an *ex-situ* conservation efforts. Some plantlets are also distributed among local people and awareness campaigns were launched lectured delivered in educational institutes and religious places for the conservation of biodiversity. Artificial ponds were also constructed and plants are grown there.

8.11.2 In-situ *Conservation*

Conservation of biodiversity in the natural habitat of a species in the form of living collections and germplasm bank where DNA clones and tissue culture techniques are carried out. Artificial regeneration of many species like *Atropa acuminata*, *Mentha arvensis*, *Colchicum luteum*, *Podophyllum hexandrum* has tried to grow under controlled conditions, and applying various agrochemicals, in the forests of Khyber Pukhton Khawa Pakistan. Many environmentalists approved that in situ conservation is better as compared to *ex-situ* if it is possible. It is essential to conserve habitats along with their biodiversity. If the natural habitats are conserved, then species will be managed on their own.

Term sircasitum conservation is also used for the conservation of biodiversity which is known to be intermediate between both conservation methods namely *ex-situ* and *in-situ*. It is linked with the traditional agricultural conservation system. They consist of the retaining wild plants when land is empty for agriculture, the growing of wild plants in home gardens, and the storage of crop seed in storerooms for later replantation.

8.12 Sustainable Development and Management of Wetlands in Pakistan

The importance of wetlands and the major threats to these areas were globally known and taken into account in the 1960s and after that the meeting for Ramsar Convention was held in 1971, keeping in view the importance and value of wetlands as well as need to protect them from invasions at a global scale. Currently, in Pakistan, many organizations worked on protecting the wetlands of the country. They have two main objectives:

1. To create and maintain the environment for sustainable and effective management of natural wetlands at the local level.
2. To implement a sustainable wetland conservation approach at representative sites that will help for successive wetland conservation.

Another organization WWF (world wildlife fund for nature) has established its center in Karachi Pakistan which aimed to conserve wetlands, enhance the biodiversity of wetlands, create awareness among the local people regarding conservation and implement the national wetland conservation strategy. In 2003 "Protection and Management of Pakistan Wetlands Project" has been started by WWF (world wildlife fund for nature) to conserve the wetlands as to conserve their biodiversity. From 2005 the implementation of this project starts practically. This project aims to support the sustainable conservation of marine and freshwater wetlands and related important biodiversity in Pakistan.

8.13 Pakistan Wetlands Conservation Policy

The proposed policy for the conservation of wetlands in Pakistan has the following objectives:

1. To protect the wetland resources land use management has been done that encourage sustainable use of natural resources. Also addressing the issues due to climate change directly affecting wetlands.
2. Coordination and collaboration have been made between agencies and sectors on wetland issues encouraged from local to international levels.
3. Wetland education, research, and data organizing should be promoted.
4. Creating awareness among the people of all levels to conserve the wetland biodiversity.
5. Initiate fundraising programs to secure the wetland managements.

8.14 Conservation of Himalayan Wetlands of Pakistan

For the conservation of wetland, the concept of their wise use has been spread widely which is defined as "the conservation of ecological characters, obtained by the application of ecosystem approaches, under the circumstances of sustainable development". It seeks to stabilize the developmental and financial requirements of the people and to regulate a healthy natural wetland ecosystem. Maintenance of the wise use of wetland is through training, research, awareness among the public, comprehensive national policies, management plans based on research which gives importance to local communities' information and traditional management systems.

Pakistan has also established a National wetlands policy which includes many wise use principles but the implementation of this policy remains patchy. Also, the conservation of the Himalayan region is limited at all government levels. While some areas in northern mountains are considered as protected areas (PAs), but the coverage percentage is less as compared to other regional countries and no Himalayan wetlands and is designated in the list of Ramsar wetlands of international importance (Khan and Arshad 2014).

On the positive side, to overcome the deficiencies of communities that depend on wetlands government introduces non-timber forest products and new cash crops. They have also introduced peat-efficient and wooden stoves, wind producing plants and biogas, check and balance on illegal timber cutting, hunting of threatened species, and awareness and training workshops for conserving these wetlands. Some lakes of the Himalayan range of Pakistan are included in protected areas and conservation measures taken in these areas.

References

Ali Z (2005) Ecology, distribution and conservation of migratory birds at Uchalli Wetlands Complex, Punjab, Pakistan. Doctoral dissertation, University of Punjab

Baloch MA, Tanık A (2008) Development of an integrated watershed management strategy for resource conservation in Balochistan Province of Pakistan. Desalination 226(1–3):38–46

Chaudhri MN, Qureshi RA (1991) Pakistan's endangered flora. 2: a checklist of rare and seriously threatened taxa of Pakistan. Pak Syst 5(1/2):1–84

Chaudhry AA (2010) Wetlands in Pakistan: what is happening to them. World Environment Day, p 5

Gill AH, Ahmad KS, Habib S, Hameed M, Ahmad MSA, Nawaz T, Ahmad F, Batool R (2012) Impact of highly saline wetland ecosystem on floral diversity of the Cholistan desert. Pak J Bot 44:107–112

Khan MH (1997) Conservation of biodiversity and endangered ecosystems in Pakistan. Documento presentado al XI Congreso Forestal Mundial

Khan AA, Arshad SANA (2014) Wetlands of Pakistan: distribution, degradation and management. Pak Geogr Rev 69(1):28–45

Khan H, Baig S (2017) High altitude wetlands of the HKH region of Northern Pakistan – status of current knowledge, challenges and research opportunities. Wetlands 37(2):371–380

Martin GJ (2014) Ethnobotany: a methods manual, vol 1. Springer, New York

Pimbert M, Gujja B, Shah M (1996) Village voices challenging wetland management policies: PRA experiences from Pakistan and India. PLA Notes 27:37–41

Qureshi NA, Ali Z (2011) Climate change, biodiversity Pakistan's scenario. J Anim Plant Sci 21(2 Suppl):358–363

Shirazi K (1993) Wetland and waterfowl conservation in Pakistan: a national perspective. Wetlands Waterfowl Conserv South West Asia:38–40

Chapter 9
Wetland Plants of Himalayas: A Case Study

FLOATING PLANTS

9.1 *Lemna minor* L.

9.1.1 Systematic Studies

Family	:	Araceae
Synonym	:	*Lenticularia vulgaris*
English name	:	Common Duckweed
Elevation	:	659 m
Life form	:	Hydrophyte, floating herb
Conservation status	:	Least concern (LC) no threats have been reported for this species.

9.1.2 General Description

It is annual herbaceous floating plant. Usually it attains height of 0.1–0.3 ft and it leaves width is approximately 1.2–1.5. Leaves are free floating on the surface of water; shape of leave is ovate to obovate or sometimes oblong also with rounded apices. Leaves are green in colour solitary in arrangement some leaves also attached with each other by a hyaline stipe. Papillae are present on apex and above node with many meshes. Roots are solitary with a wingless sheath, cap of root is obtuse and having lateral floral pocket (Fig. 9.1a–b).

M. A. Butt et al., *Wetland Plants*, https://doi.org/10.1007/978-3-030-69258-2_9

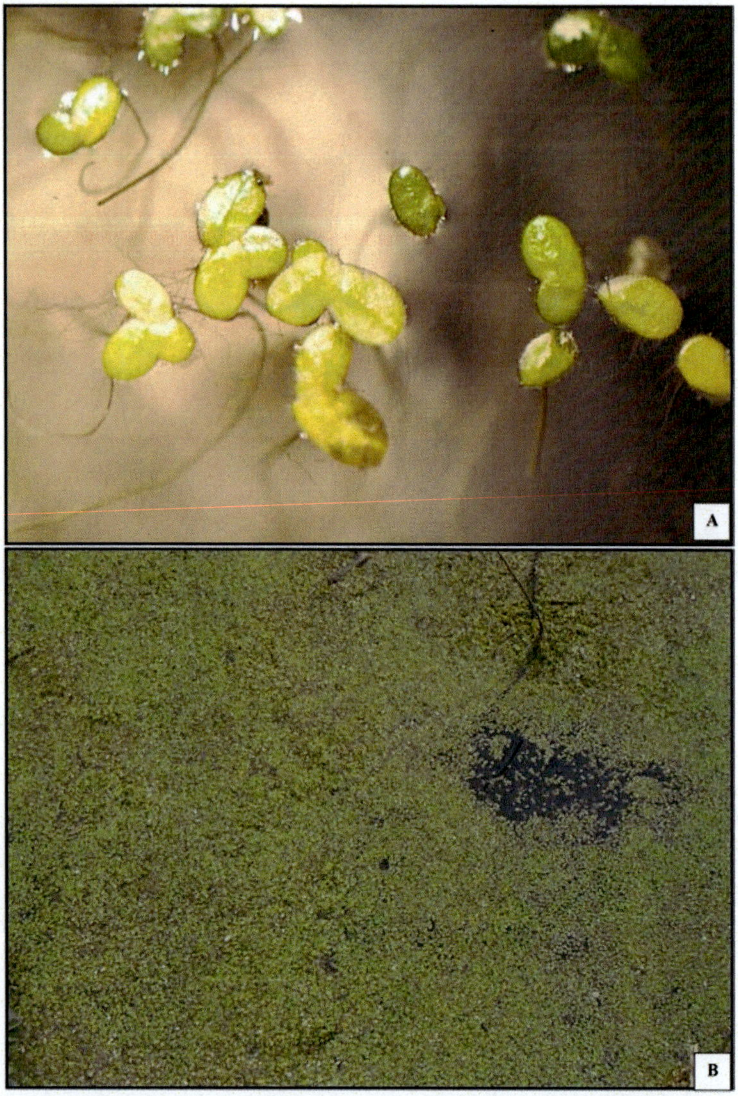

Fig. 9.1 (**a–b**) Field photograph of *Lemna minor*

9.1.3 Distribution

It is common plant and found in Pakistan, temperate Eurasia, Iran, India, Afghanistan, Africa, North America and New Zealand.

9.1.4 Phenological Characters

Flowers are rarely seen. They are tiny and white in colour.

9.1.5 Ecological Characters

This plant grows in aquatic environment in freshwater ponds, stagnant water, slow-moving streams and in lakes.

9.1.6 Medicinal Uses and Other Uses

Whole plant is used in making medicine which is a good remedy for asthma. Decoction of leaves is used during liver disorders that cause yellowing of skin. Also decoction is good for arthritis patients. Externally this plant is used on skin to cure allergies and is also use for washing eyes. Dried leaves are used to repel the mosquitos. This plant is also used in the treatment of waste water as it absorbs nutrients from the surface of water. It is also cooked as vegetable.

9.2 *Nymphaea nouchali* Burm.f.

9.2.1 Systematic Studies

Family	:	Nymphaeaceae
Synonym	:	*Castalia stellaris* Salisb.,
English name	:	Blue lotus
Elevation	:	590 m
Life form	:	Hydrophyte, floating herb
Conservation status	:	Least concern (LC) no threats have been reported for this species.

9.2.2 General Description

It is a perennial herb, includes in floating plants, leaves are 9–16 cm in length, elliptical in appearance, margins are either entire or slightly dentate dull reddish from the upper surface. Pedicle and petiole are very long. Flowers are bright blue in colour, scented, Sepals are green in colour with entire margins. Petals are bright in colour, linear to lanceolate in shape that are arranged in rows. Stamens are present with the appearance of blue appendages with stigmatic rays. Aril fleshy seeds are also present (Fig. 9.2a–b).

Fig. 9.2 (**a–b**) Field photograph of *Nymphaea nouchali*

9.2.3 Distribution

It is distributed among tropical Africa, found in Himalayan regions of Pakistan, India and Nepal, also found in South Andaman and Bangladesh.

9.2.4 Phenological Characters

The flowering and fruiting of this plant is observed in May to August and best collection season for seeds are from July to August.

9.2.5 Ecological Characters

This species is mostly found in stagnant water, lakes and ponds. This plant is now a day used as an ornamental plant.

9.2.6 Medicinal and Edible Uses

It contains the active alkaloids which is sedative and an aphrodisiac. The root of the plant was used by monks and nuns for hundreds of years as an aphrodisiac, being crushed and mixed with wine. A decoction of the root is used in the treatment of dysentery or diarrhea. Externally it can be used to make a douche to treat vaginal soreness or discharges. A complete cure of uterine cancer by a decoction and uterine injection has been recorded.

9.3 *Nymphaea alba* L.

9.3.1 Systematic Studies

Family	:	Nymphaeaceae
Synonym	:	*Castalia alba*
English name	:	Water lily
Elevation	:	590 m
Life form	:	Hydrophytes, floating herb
Conservation status	:	Least Concern (LC) no threat have been reported for this species.

9.3.2 General Description

It is a perennial herb having rhizome on horizontal side. Leaves have long petioles, floating on the surface of water, sub-orbicular and cordate in appearance. The size of leaves ranges from 11 to 26 cm with divergent lobes. Flowers are 13–22 cm in size sometimes they are scented while rarely they have any fragrance. Sepals are oblong in shape sometimes they are also lanceolate, white-yellowish green in colour and obscurely veined. Petals are many in number, white or pink in colour. Stamens are numerous and arranged in 3 rows, filaments are broad as anthers. Stigmas flat, rays 7–20. Fruit is globose, green in colour, seeds many and ellipsoid in appearance (Fig. 9.3a–b).

9.3.3 Distribution

This is common plant distributed in Europe and Asia (Turkey, Iran and Pakistan and Kashmir).

9.3.4 Phenological Characters

The plant blooms in the month of June to August.

9.3.5 Ecological Characters

This plant is cultivated and usually used as ornamental purposes. Mostly found growing in ponds, lakes, stagnant water and in swimming pools.

9.3.6 Medicinal and Edible Uses

The rhizome of this plant is highly medicinal and it is used in making medicines for anti-scrofulatic, astringent, cardio-tonic, demulcent and sedative drugs. The decoction of rhizome is taken two times a day to cure dysentery or diarrhea. Decoction of roots is also taken in lungs disorders and it is used to gargle during sore throat. Externally it can be used to make a douche to treat vaginal soreness or discharges. Seed of this plant is used as a substitute for coffee.

Fig. 9.3 (**a–b**) Field photograph of *Nymphea alba*

9.4 *Potamogeton nodosus* Poir.

9.4.1 Systematic Studies

Family	:	Potamogetonaceae
Synonym	:	*Potamogeton americanus*
English name	:	Longleaf pondweed
Elevation	:	679 m
Life form	:	Hydrophytes, floating herb
Conservation status	:	Least concern (LC) no threats have been found to this species.

9.4.2 General Description

This plant is perennial rhizome bearing aquatic herb. Usually floats on the surface of water. Its stem is branched, leafy and simple. Leaves are heterophyllous, leaves are sometimes submerged having petiolate, lanceolate and the veins are broad and more than 9 in number, thin and translucent. Floating leaves are ovate in shape and their margins are entire. Stipules are also present which are free, keeled and lanceolate in shape. Inflorescence is in the form of spikes. Flowers are arranged in whorls. Perianth is segmented and small, obtuse in shape with entire margins. Stigma is short. Fruits are obliquely obovoid, spongy and shortly beaked (Fig. 9.4a–b).

9.4.3 Distribution

This plant is commonly distributed all over the world in warmer regions including Europe, North and Central America, Africa and Asia. In Asia this plant is found in Indo- Pak subcontinent, Indonesia, Japan, Kazakhstan, Myanmar, Nepal, New Guinea, Russia, Sri Lanka, Tajikistan, Thailand, Turkmenistan, Uzbekistan and Vietnam.

9.4.4 Phenological Characters

The flowering season starts from April and ends in August. Fruits appears on plant in the month of August to September. After fruits the plant disappears in winter and regrowth of plant occurs in April.

Fig. 9.4 (**a–b**) Field photograph of *Potamegeton nodusus*

9.4.5 Ecological Characters

It is found growing in ponds, lakes, stagnant water and water which is slightly alkaline in nature.

9.4.6 Medicinal Uses

Leaves of this plant are used in making medicines. Fresh leaves are crushed and form paste from them and applied on body as pain balm. Leaf decoction is also used in urine detention in humans. It is also given to the livestock to relief them from fever. Leaves extract of this plant is also used to treat chills. This plant is also used in treating liver cancer, leaves paste are also applied on skin to cure acne and pimples. Decoction of rhizome is taken to treat respiratory disorders like common cold and cough. Leaves paste is also applied on wounds for quick healing. Infusion of leaves is taken to cure abdominal discomfort among humans.

9.4.7 Toxicity

Excessive growth of this plant on lakes and ponds can cause toxicity to water. Excessive intake of this plant by cattle can cause diarrhea.

9.5 *Marsilea quadrifolia* L.

9.5.1 Systematic Study

Family	:	Marsileaceae
Synonym	:	–
English name	:	Water clover
Elevation	:	622 m
Life form	:	Pteridophytes herb
Conservation status		Least Concern (LC) no major threats have been reported for this species.

9.5.2 Botanical Description

This plant is annual herb including in pteridophytes group. The rhizome of this plant is creepy and bears internodes which are of 9.9 cm in size. Roots arise from internodes or sometimes from nodes also, hoots formed at nodes. The surface of rhizome is glabrate. Fronds are light green to straw in colour, 10.2–19 cm in length, glabrate in surface; pinnae is triangular or fan shaped, 0.9 to 1.5 × 0.5 to 1.7 cm in size, margins of fronds are entire. Sporocarps are attached at base of stipe where stipe joins

Fig. 9.5 (**a–b**) Field photograph of *Marsilea quadrifolia*

rhizome, sometimes they become persistent after wilting of fronds after this they again arise from rhizome directly. Peduncles are also present having 5.3–12.6 mm thickness. The surface of sporocarp is glabrate, sporocarp is dark brown to black in appearance, elliptic in shape from lateral view (Fig. 9.5a–b).

9.5.3 Distribution

This plant is annual and is cosmopolitan in distribution mostly found in Europe (Central and Southern), Asia (Afghanistan, south-west India, China, Japan, Pakistan and Vietnam).

9.5.4 Phenology

It is pteridophytes so they do not produce flowers. Spores are produce in sporocarp in the month of July to September.

9.5.5 Ecology

Mostly found grown in ponds, stagnant water, paddy field, ditches, shallow water of lakes, quiet sections of rivers and streams and on wet shores.

9.5.6 Medicinal Uses

Leaves are important part used as medicinal purpose as the juice is used to treat liver disorders and febrifuge. Paste of the leaves is applied on the wounds to relieve inflammation.

9.6 *Azolla pinnata* R. Br.

9.6.1 Systematic Study

Family	:	Salviniaceae
Synonym	:	*Salvinia imbricata* Roxb. ex Griff.
English name	:	Mosquito fern
Elevation	:	331 m
Life form	:	Pteridophytes herb
Conservation status	:	Listed as Least Concern as it is a widespread species and faces no major threats

9.6.2 Botanical Description

This plant is ovate or triangular in appearance. Stem is slender like and lateral branches of stem is creeping, dichotomous, having roots growth downwards. Fronds are small just like sesame seeds, alternate in arrangement, blade of trapezoidal separated into 2 lobes i.e. dorsal and ventral. The dorsal lobes are green in colour and they are floating, during autumn season they turn into purple colour also they are oblong in shape, and margins are colorless. The ventral lobes are submersed in water and appear in red colour, or sometimes they are also colorless and transparent. Sporocarps are formed in pairs; megasporocarp and microsporocarps; megasporocarps are smaller in size, with ovate appearance, and having beaked apex. Microsporocarps are large in size, globose or sometimes peach-shaped, apex having beak which is short, microsporangia are long stalked, having 65 microspores embedded in 5–8 massulae which have silk like hairs on its surface (Fig. 9.6a–b).

9.6.3 Distribution

This plant is distributed in tropical regions of Asia (South East Asia, South East China, Korea, Philippines and South Japan), Australia and in some countries of Africa.

9.6.4 Phenology

This plant grows throughout the year as it needs aquatic environment to grow so it is common in tropical and sub-tropical areas. It produces sporangia that bearing spores in the month of June to September.

9.6.5 Ecology

As this plant is floating fern it is found in ponds, ditches, wet areas of lakes, stagnant water, rice fields and in backwater having no waves.

Fig. 9.6 (**a–b**) Field photograph of *Azolla pinnata*

9.6.6 Medicinal and Edible Uses

The frond of this plant is used to induce perspiration and other skin related diseases. Fronds of this plant are also used by humans as salads, omelette and making mungo. This can also be added to flour to increase the fiber content in the meal. It also acts as bio-fertilizers in paddy field. It is also known that this plant is rich with vitamins and minerals so it is given to broiler chicken to increase their weight and also increases the production of eggs. As it is rich source of proteins so it is given to fishes as their feed.

9.6.7 Toxicity

It can cause toxicity to water as it forms mat on the surface of water and thus lowers the percentage of oxygen.

9.7 *Spirodela polyrrhiza* (L.) Schleid

9.7.1 Systematic Study

Family	:	Araceae
Synonym	:	*Lemna major* Griff.
English name	:	Greater duckweed
Elevation	:	331 m
Life form	:	Floating herb, hydrophytes
Conservation status	:	Listed as Least Concern as it is a widespread species and faces no major threats

9.7.2 Botanical Description

It is a floating aquatic herb. Leaves are cohering in groups of 2–5 and sometimes solitary in arrangement, 9.3 to 10.1 × 3.9 to 8.6 mm in size ovate to orbicular in shape, having obtuse shape from base, apex of these leaves are rounded or sometimes obtuse; upper side of the leaf is flat shiny, smooth and green in color while lower side is rough, flat and brownish red in colour. Plant reproduces asexually through budding that arise from the root insertion point of parent leaf. Roots originated from the thick part of leaves and they are 5–9 in numbers; root also contains

Fig. 9.7 (**a–b**) Field photograph of *Spirodela polyrhiza*

cap which is acute. Spathe covers the flowers with slit-pouch each flower contains one stamen. Ovary is having 1 or 2 ovules. Fruit is 1 to 2 seeded, winged and having smooth surface (Fig. 9.7a–b).

9.7.3 Distribution

This plant is commonly distributed throughout the world including Europe, North Africa, Asia (Pakistan, India, China and Nepal), America and Australia.

9.7.4 Phenology

Flowers of this plant blooms in September to November.

9.7.5 Ecology

This plant is found in warm environmental condition in fresh lentic waters. It also found growing in lakes, ponds, dams, stagnant water, in reeds, sewage canals, and slow moving water of rivers and also found growing with *Lemna*, *Wolffia*, and *Pistia* species.

9.7.6 Medicinal and Edible Uses

The whole plant is medicinal and its extract is useful in conditions like antipruritic, cardio tonic, antipyretic, diaphoretic, carminative and diuretic. Extract of whole plant is used in treatment of common colds, measles and difficulty during urination. Paste of the leaves are also good in treating skin disorders like rashes, burns, sunburn and also cure pimples on skin. This plant is also eaten by ducks and fishes as a source of food. It is also used to purify water as it has capacity to extract toxic materials from water and purify it. It is edible and in some counties it is also eaten as salad. Leaves of this plant are crushed and wrap in cotton cloth put on the area of sun burn to relief pitching and pain. It is also a good remedy for inflammation in respiratory tract and good in rheumatoid arthritis. Poultice of leaves wrapped on insect bite to relieve pain. Juice of this plant is also good in treating urinary tract disorders.

9.7.7 Toxicity

Excess amount of intake of this plant extract can stimulate the rashes of measles. It contains high amount of calcium oxalate which can cause kidney stones so huge consumption should not be taken.

EMERGENTS

9.8 *Sagittaria trifolia* L.

9.8.1 *Systematic Study*

Family	:	Alismataceae
Synonym	:	*Sagitta aquatica* (Lam.)
English name	:	Arrowhead
Elevation	:	1981 m
Life form	:	Perennial herb, emergent
Conservation status	:	Least concern (LC) no reports have been found about threating of this species.

9.8.2 *Botanical Description*

This plant bears tubers on the tip of thick stolons. Leaves are mostly emerged in water, sagittate in arrangement but sometimes they seems to be hastate, leaf blade is ovate to lanceolate in shape, 8.2–17.9 cm long along the midrib and 5.5–7.2 cm broad, apex is acute, triangular basal lobes, these lobes are longer than blade, these lobes are sharply acute from tip; petiole is 60.9–75.2 cm in length and also triangular in shape. Inflorescence is raceme which is 29–51 cm long. Flowers are formed in the form of 2–6 whorls; they are unisexual, white in colour; length of pedicels is 8.2–15.3 mm, this size varies as they are short in females; bracts are also present which are triangular or sometimes ovate in shape. Sepals are elliptic in shape, margins are membranous. Petals white in colour with elliptic in appearance. Long stamens with yellowish anthers and oblong filaments are present. Fruits are in the form of achene which is triangularly obovate in shape (Fig. 9.8a–b).

9.8.3 *Distribution*

This plant is distributed throughout the world specifically it is found in Asia (Iraq, Iran and West Pakistan, China, Malaysia and Japan), Australia, Europe (Hawaii).

Fig. 9.8 (**a–b**) Field photographs of *Sagittaria trifolia*

9.8.4 Phenology

This plant blooms in the month of April to September and fruiting period starts from September to November.

9.8.5 Ecology

This plant found growing along ponds, banks of lakes, marshes, paddy fields and irrigation channels. Some species of this genus is also cultivated for ornamental purposes in gardens and lawns.

9.8.6 Medicinal and Edible Uses

The tuber of this plant is highly medicinal and is used in discutient, galactofuge and may induce premature birth. Whole plant is antiscorbutic and act as diuretic. Leaves are crushed with water to make thick paste and it is applied on skin to cure skin problems. Plant is also cultivated as ornamental plant. Corm of this plant is also used in making soup in some countries. Dried corms are ground to powder and used as porridge and also in making breads. Leaves are also used as vegetable in some countries.

9.8.7 Toxicity

Skin of tubers of this plant is bitter in taste and it is reported to be highly toxic.

9.9 *Schoenoplectus lacustris* (L.) Palla

9.9.1 Systematic study

Family	:	Cyperaceae
Synonym	:	*Cyperus americanus* Garsault
English name	:	lakeshore bulrush
Elevation	:	659 m
Life form	:	Emergent herb
Conservation status	:	Least Concern, population stable

Fig. 9.9 (**a–b**) Field photograph of *Schoenoplectus lacustris*

9.9.2 Botanical Description

It is a perennial herb usually attains height of 90–90 cm having horizontal rhizomes. Stem is soft, terete with scarious margins. Inflorescence of this plant is spike and 51–80 mm long, lowest bract is thick, erect and also terete with acute apex. Perianth is 5–8 in number, brownish red in colour; stamens are more than 2 in number, anthers are reddish brown, having connective tip; 2–3 stigmas can be observed. Fruit is in the form of nut, which is either obovoid or triangular, plano-convex, surface of which is finely reticulate (Fig. 9.9a–b).

9.9.3 Distribution

This plant is common so it spread throughout the world in Europe, East and Central Asia, Africa and south to North America. In Asia this plant is reported from Pakistan, Kashmir, Afghanistan, China, India, Nepal, Iran and Turkey.

9.9.4 Phenology

The booming season for this plant starts from June to August, and the seeds ripen from August to September.

9.9.5 Ecology

Grown on wet soil especially at bogs, rivers and lakes, pond margins, silt dominant areas and in calcareous conditions.

9.9.6 Medicinal and edible Uses

As this plant is rich source of starch so it is used either raw or cooked. Buds present at the end of rhizomes are sweet in taste so they are eaten raw. Powder of dried seeds is used in making bakery products. Roots of this plant is converted into decoction on boiling with water and taken to cure constipation and also helpful in urine blockage issues.

9.10 *Cyperus alternifolius* L.

9.10.1 Systematic Study

Family	:	Cyperaceae
Synonym	:	*Cyperus onustus* Steud.
English name	:	Umbrella sedge
Elevation	:	682 m
Life form	:	Perennial herb
Conservation status	:	Least Concern, Population increasing

9.10.2 Botanical Description

It is tufted perennial herb usually attained height of 30–95 cm with short and horizontal rhizome. Stem surface is smooth and scabrous. Blades of leaf are reduced to form sheath which is yellow from upper part and dark brown from lower part, surface of leaves contains large lines on it. Inflorescence is yellowish green in colour and compound anthelodium in arrangement with acute apex, primary anthelodia is green but sometimes in maturity it turns brown with spongy base. Secondary anthelodia having bracts and tertiary anthelodia have cluster of spikes. Stamens are usually 3 in number. Fruit is in the form of nuts. Shape of nuts is ellipsoid or sometimes obovoid also. Light brown in colour and finely papillose (Fig. 9.10a–b).

9.10.3 Distribution

This plant is found reported in few countries in Madagascar, Mauritius, Pakistan, China, Nepal and Kashmir.

9.10.4 Phenology

This plant starts flowers from May to June and fruits in June to July.

9.10.5 Ecology

This species grows in swamps, wet grasslands and beside streams, in stagnant water, sewage canals. It grows in slight sandy soil near water. This plant can tolerate flooded conditions when the water level rises. It only grows in bright sunlight in dark places growth of shoots retarded.

9.10.6 Medicinal and Edible Uses

Dried stems of this plant are used in making bottom of wooden chairs. Pulp of this plant is processed in industries and helps in making paper. In some countries this plant is also grown as ornamental purposes. It is also used in making forage for livestock's in some countries. As this plant contains active secondary metabolite alkaloid so the decoction made from boiling whole plant in water is used to treat digestive disorder and also helps in curing different types of fever and to disinfect

Fig. 9.10 (**a–b**) Field photograph of *Cyperus alternifolius*

wounds from germs ash of this plant is applied as a poultice on them. In some areas this plant is used in religious customs during funeral activities. It is also used as superstitious rituals to cure sickness.

9.10.7 Toxicity

Excessive intake of this plant can cause minor illness as this plant is less toxic to humans but it may cause diarrhea, dysentery and sometimes vomiting and dizziness also.

9.11 *Juncus articulatus* L.

9.11.1 Systematic Study

Family	:	Juncaceae
Synonym	:	*Juncus affinis* Gaudin
English name	:	Joint leaf rush
Elevation	:	590 m
Life form	:	Perennial herb
Conservation status	:	Least Concern, Population stable

9.11.2 Botanical Description

This plant is perennial, having short rhizome which is green in colour. This plant attains height of 15.2–90 cm. plant is erect but sometimes it seems to be prostate and roots originated from the nodes which are present on the lower side. Stem is terete with cauline leaves on it. Stem is linear with short leaves on it, green in colour, on the base sheath is attach which is brown in colour. Inflorescence is terminal, branched dichotomously, with many stalked heads. Flowers are pale yellow to dark brown in colour, sessile with lanceolate perianth. Perianth are usually bigger than stamens, stamens are 6 in number having anthers which are equal in size to filaments. Fruits are in the form of capsule which is broad, oblong-ovoid in shape, brown or green in colour, seeds are ovoid to ellipsoidal in shape and dark brown in colour (Fig. 9.11a–b).

9.11.3 Distribution

This plant is distributed commonly in almost all continents of world including Europe, North America, N. Africa, Central and S.W. Asia, to Pakistan and the Himalayas and Australia.

Fig. 9.11 (**a–b**) Field photograph *of Juncus articulatus*

9.11.4 Phenology

The blooming season of this plant starts from June and ends in July, while they produce fruits at the end of July to September.

9.11.5 Ecology

This plant prefers wet lands and wet soil conditions to grow. It prefers growing in fresh water streams, wet banks of rivers, lakes, ponds, irrigation channels also grows in wet mud or sand in shallow, still or slow-moving water. Grow in medium (loamy), light (sandy), and heavy (clay) soils. In bright sunlight the plant starts wilting it grow in shaded and semi shaded areas near water courses. It can bear strong winds but not maritime exposure.

9.11.6 Medicinal and Edible Uses

Leaves of this plant is used in making herbal teas which helps in curing chronic constipation, jaundice, throat infections and urinary tract disorders. Root powder is good treatment for dysentery and is good in constipation also. Whole plant decoction is taken in insomnia as it has sedative properties also. Stem of this plant is dried and used to make ropes. Shoots of this plant is used in making baskets, thatching and weaving mats. Peeled stems are used in making candles. Fibers extracted from stems are used in making paper. Fiber of this plant is mixed with mulberry fibers and used in making pencils. Also cultivated as an ornamental plant in gardens.

9.11.7 Toxicity

If this plant is taken in excessive amount by mammals it can cause diarrhea and dysentery in them.

9.12 *Eclipta prostrata* (L.) L.

9.12.1 Systematic Study

Family	:	Compositae
Synonym	:	*Acmella lanceolata* Link ex Spreng.
English name	:	False daisy
Elevation	:	Below 1600 m
Life form	:	Annual herb
Conservation status	:	Least concern

9.12.2 Botanical Description

A prostrate or reclining to erect, often branched, annual or perennial herb, 30–100 cm tall. Stem is cylindrical, green or purplish, rooting at basal nodes, and often covered with long white hairs. Leaf is oblong to lance-shaped, opposite, sessile or short-stalked, with more or less coarse hairs margins entire or slightly toothed, up to 2–16 cm long. Inflorescence is terminal and axillary, about 1 cm across, white or cream, on peduncles to 7 cm long. Fruit is achene, densely warted, either brown or black, 2–3 mm long (Fig. 9.12a–b).

9.12.3 Distribution

This plant is distributed throughout the world in Asia, Africa, Australia and Europe. In Pakistan it is found in areas Mingora, Shangla, Mansehra, Islamabad, Azad Kashmir and Peshawar.

9.12.4 Phenology

It starts flowering from the month of September to November and fruits are seen on this plant in the month of November.

9.12.5 Ecology

Plant need aquatic environment to survive as it is mostly found grown in the banks of pools and lakes, on the edge of rivers, swamps, channels of irrigation, streams and ditches.

9.12.6 Medicinal and Edible Uses

Paste of leaves are mixed with butter and used to cure jaundice. Juice extracted from whole plant is used to treat urinary tract infections. Leaf powder of this plant is mixed with coconut oil which helps to treat hair loss. Juice of leaves is used to treat respiratory disorders. Paste obtained by crushing the leaves of this plant is used to relief pain during scorpion sting and also act as antidote for reptiles or insect bites. Blue colored dye is obtained from the stem of this plant which is used in hair dying or as ink for tattoos making on the body.

Fig. 9.12 (**a–b**) Field photograph of *Eclipta prostata*

9.13 *Eichhornia crassipes* (Mart.) Solms

9.13.1 *Systematic Studies*

Family	:	Pontederiaceae
Synonym	:	*Eichhornia cordifolia* Gand
English name	:	Common water hyacinth
Elevation	:	200–500 m
Life form	:	Annual herb

9.13.2 *Botanical Description*

Herb, usually free-floating but occasionally rooted in mud, 9–90 cm. Leaves 9–8, clustered on short rhizomes, erect, very variable; in free-floating plants: petiole 5–10 cm, markedly swollen below; lamina 9–9 cm wide, ovate to circular; in older plants rooted in mud, the petiole is longer and not swollen. Inflorescence a c.8-flowered showy spike, raised above leaves. Perianth tube 20 mm curved; limb 5–7 cm in diameter, slightly 2-lipped, pale mauve with the central lobe broadest with large blue area and central yellow spot (Fig. 9.13a–b).

9.13.3 *Distribution*

This plant is distributed in South and Central America and Asia. In Pakistan it is distributed in areas Swabi, Peshawar, Mardan, Bannu, Sialkot, DI Khan.

9.13.4 *Phenology*

This plant flowers in the month of July to September and fruiting season starts from August to September.

9.13.5 *Ecology*

It grows in shallow temporary ponds, wetlands and marshes, sluggish flowing waters and large lakes, reservoirs, and rivers.

Fig. 9.13 **(a–b)** Field photograph of *Eichhornia crassipes*

9.13.6 Medicinal and Edible Uses

The whole plant is used as fodder for livestock because it can cause fattening of them. In some areas of the world this plant leaves and petioles are cooked and used as vegetable. Flowers are used for mediating horses and the rhizome is also used to make different types of tonic. On agricultural point of view this plant is also used as green manure. Very little amount of this plant is also used in paper making. This plant is also known to be used for the waste water treatment. It is also used as fuel in dried form in some rural communities. Leaves and flowers of this plant is also used in making baskets.

9.14 *Phyla nodiflora* (L.) Greene

9.14.1 Systematic Study

Family	:	Verbenaceae
Synonym	:	*Blairia nodiflora* (L.) Gaertn.
English name	:	Turkey tangle frog fruit
Elevation	:	300–2300 m
Life form	:	Perennial herb
Conservation status	:	Least Concern Population stable

9.14.2 Botanical Description

A perennial herb found in wet places. Leaves are round in shape, spatulate, opposite, simple with serrate (above) and entire (below) margins, acuminate apex, attenuate base, pinnate venation and scrufy surface having small hairs on it. Length of leaves is 1.3–2.9 cm and its width is 0.7–0.8 cm. The stem is purple to green in colour; surface of stem is glabrous and the stem is herbaceous. Roots are originated from leaf axils and this plant contains taproot. Sometimes fibrous roots also originated from the nodes of stem. Flowers are purple, white and light pink in colour which is arranged in a spike inflorescence. Fruit is ovate, brown in color and form in capsule (Fig. 9.14a–b).

Fig. 9.14 **(a–b)** Field photograph of *Phyla nodiflora*

9.14.3 Distribution

This plant is distributed in sub-tropical and tropical regions of the world.

9.14.4 Phenology

This plant flowers throughout the year and fruits of this plant originate after flowering.

9.14.5 Ecology

It is mostly found growing in the channels of water, streams, lakes, bank of rivers, wet muddy places, in irrigation channel and moist banks of ponds.

9.14.6 Medicinal and Edible uses

Poultice made from the leaves are applied on boils for soothing effect. Paste of leaves is applied externally for curing burns, ulcers, wounds and facial skin problems. Roots are grinded and mixed with water to form juice which is used in gastric troubles. Dried leaves of this plant also help in making tea. Also used in making lawns and used as alternative for grass.

9.14.7 Toxicity

Excessive amount of intake of this plant can cause live and nervous system damage.

9.15 *Eleocharis palustris* (L.) Roem. & Schult

9.15.1 Systematic Study

Family	:	Cyperaceae
Synonym	:	*Bulbostylis palustris* (L.) Steven
English name	:	Common spike-rush
Elevation	:	500–3000 m
Life form	:	Perennial herb
Conservation status	:	Least Concern

9.15.2 Botanical Description

This plant is perennial herb usually attains height of 31–90 cm. size of stem ranges between 1.6–2.7 mm, grooved having veins present on both sides. Sheath is present instead of leaves, from which lower sheath is dark brown in colour while the upper sheath is reddish brown in colour. Inflorescence is in the form of spike which is 11–23 cm in length and 3.9–5.1 mm broad, flowers containing glums which are 3.1–3.6 mm in size. Upper part of glume is black to dark brown in colour. Anthers are 1.6–2.1 mm in size. Fruit is in the form of nut which is1.9–1.5 mm in length and 1.1–1.9 mm in width. Margins of the fruit are slightly protruding, dark to pale brown in colour (Fig. 9.15a–b).

9.15.3 Distribution

This plant is distributed in Himalayan regions in Asia and also reported from Europe and Afghanistan. In Pakistan this plant is distributed in higher altitude areas like Gilgit- Baltistan, Chitral, and Skardu.

9.15.4 Phenology

Flowering period of this plant starts from May and ends in the month of July.

9.15.5 Ecology

It needs aquatic environment to flourish, mostly found growing in marshes, swamps, bogs, wet banks of rivers and lakes sometimes also found in stagnant water.

9.15.6 Medicinal and edible uses

Plant has been used for stuffing pillows making, sitting pads and also used for bedding. This plant is also used for waste water treatment. Leaves of this plant are rich in high protein and fibers so it is eaten by livestock and easily digestible for them. Ducks and geese eat the flowers and also used the stems of this plant for making nests.

Fig. 9.15 (**a–b**) Field photograph of *Eleocharis palustris*

9.15.7 Toxicity

Eating this plant is toxic as it may cause skin allergies and eyes irritations.

SUBMEREGED PLANTS

9.16 *Vallisneria spiralis* L.

9.16.1 Systematic Study

Family	:	Hydrocharitaceae
Synonym	:	*Vallisneria aethiopica* Fenzl
English name	:	Eel grass
Elevation	:	563 m
Life form	:	Perennial herb
Conservation status	:	Least Concern

9.16.2 Botanical Description

This plant is totally submerged with 11 mm in length, having stolons which are terete. Shape of leaf is oblong but sometimes linear leaves are also present, ribbon type, transparent to slightly green in colour and having parallel costate. Margins of leaves are denticulate but sometimes entire also. Male spathes are oblong in appearance; peduncles are also seen having size 2.2–3.1 cm lengthwise. Bursting of male spathes lubricates the male flowers which are pedicellate, sepals are unequal in sizes, and stamens are solitary or 2 in number. Female spathe enclosed the ovary which is bi-lobed at its apex, filiform peduncle; spirally coiled fruit is covered by long leaves. Female flower are solitary in arrangement; sepals are ovate to oblong in shape. Stigmas are hairy. Fruit is yellowish green in colour; seeds are many in number and cylindrical in shape (Fig. 9.16a–b).

9.16.3 Distribution

It is distributed in temperate, tropical and subtropical regions of the world.

Fig. 9.16 (**a–b**) Field photograph of *Vallisneria spiralis*

9.16.4 Phenology

Flowering season of this plant starts from October and ends in the month of March. After flowering fruiting starts from March to April.

9.16.5 Ecology

This plant is found growing in stagnant water, streams, slow moving water in rivers, lakes, irrigation channels and in ponds.

9.16.6 Medicinal and Edible uses

Decoction made by boiling whole plant in water to treat stomach troubles and also helpful in curing women gynecological disorders. Leucorrhoea in women is treated by using decoction of this plant. To improve the appetite tea of dried leaves of this plant along with *Sesamum indicum* is taken.

9.17 *Potamogeton lucens* L.

9.17.1 Systematic Study

Family	:	Potamogetonaceae
Synonym	:	*Buccaferrea lucida* Bubani
English name	:	Shining pondweed
Elevation	:	595 m
Life form	:	Perennial herb
Conservation status	:	Least Concern (Population stable)

9.17.2 Botanical Description

This plant is rhizomatous hydrophyte usually grows from 6.2 cm to 9 cm in length having branched stem. Leaves are sometimes petiolate and sometimes they are sub-sessile also. Leaves are transparent to brownish green in colour, veins are present in network, and margins are slightly dentate, lanceolate in arrangement and obovate in shape. Size of leaf ranging between 6.5 and 8 cm and width of leaf is 2.1–3.9 cm, apex is acuminate or acute. Stipules are free, arrange as lanceolate, sub-obtuse in shape, 2-keeled. Inflorescence is spike, cylindrical in shape and dense. Fruits are shortly beaked, round in shape, beak is decurrent and central (Fig. 9.17a–b).

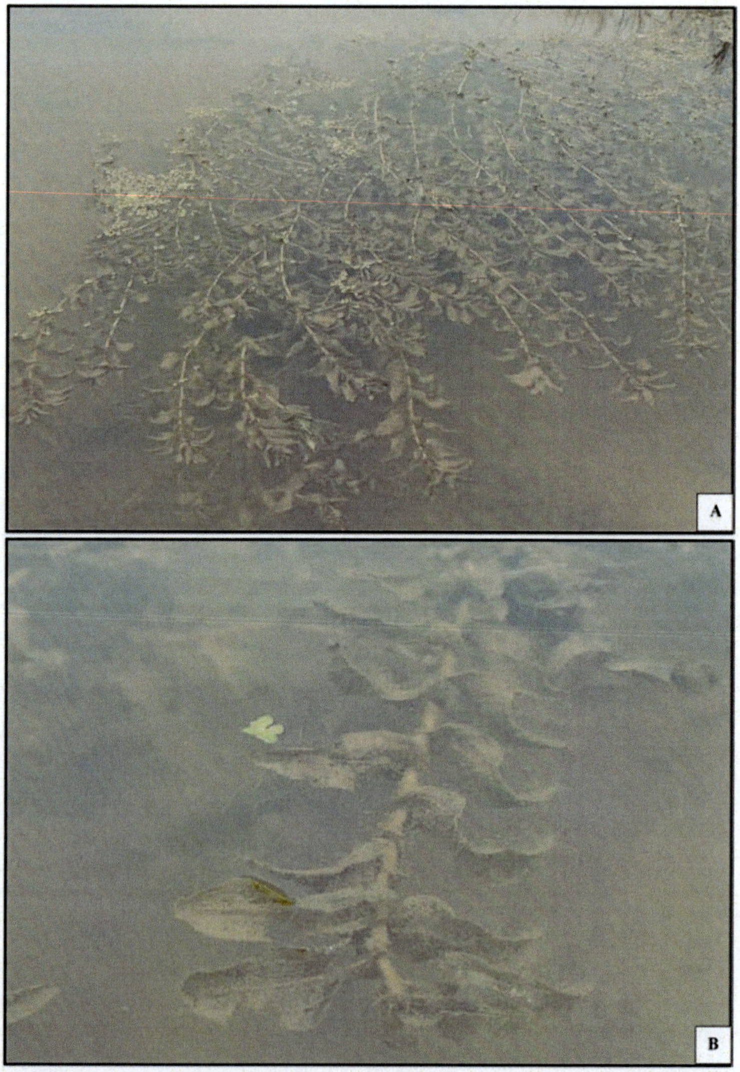

Fig. 9.17 **(a–b)** Field photographs of *Potamogeton lucens*

9.17.3 Distribution

It is distributed in different countries around the world including most countries of Asia; Uzbekistan, India, Pakistan, Nepal, China, Kazakhstan, Kyrgyzstan, Myanmar, Afghanistan, Philippines, Tajikistan, Russia and Turkmenistan.

9.17.4 Phenology

Blooming season starts from July and ends in August. After that fruits arise in the month of August.

9.17.5 Ecology

Plant is found growing in shallow water, stagnant water, sewage tanks, shores of sea, irrigation channels, and slow moving water of river, streams and ponds.

9.17.6 Medicinal and Edible uses

Roots of this plant are eaten raw as it is sweet in taste. In some parts of world this plant is also used as vegetable due to presence of nutrients. Medicinally this plant acts as resolvent and febrifuge.

9.17.7 Toxicity

No toxic effect has been reported for this plant.

MOIST PLACES PLANTS

9.18 *Rhodiola coccinea* (Royle) Boriss.

9.18.1 Systematic Study

Family	:	Crassulaceae
Synonym	:	*Sedum coccineum* Royle
English name	:	Rosy evening-primrose
Elevation	:	3657 m
Life form	:	Perennial herb
Conservation status	:	Endangered specie

9.18.2 Botanical Description

Root is long and cylindrical. Rhizome is also present which is thick and elongated with 1–3 cm broad. Leaves are scaly and originated from the apical part of rhizome. Each rhizome bears 3–9 flowering branches which are simple, glabrous and erect. Flowers are brownish or red in colour smooth to papillae, usually attains height of 1–5 cm and 0.3–2 mm broad at anthesis. Flowers are unisexual and pedicellate. Calyx is lobed which are basally connate, glabrous and triangular to oblong in appearance. Petals were having entire margins, glabrous, oblong to obovate. Stamens are 7–11 in number longer same as size of petals, arranged in two whorls, basifixed and epipetalous. Scales of nectar are oblong. Seeds are 5–13 in each fruit, ellipsoidal in appearance, brown in colour and terminally winged (Fig. 9.18a–b).

9.18.3 Distribution

Kashmir; District Neelum (Ratti Gali, Chita Khatta), Pakistan; Gilgit (Deosai), China, Nepal. Endemic to Himalayan region

9.18.4 Phenology

Flowering season starts in June and ends in July.

Fig. 9.18 (**a–b**) Field photograph of *Rhodiola coccinea*

9.18.5 Ecology

Found growing in moist and wet places, near lakes and streams.

9.18.6 Medicinal and Edible Uses

It is a wild edible plant. Leaves of this plant is cooked and eaten as vegetable. In some regions rhizome is also taken as a food. Roots are boiled in water to soften them and applied on the joints to relieve pain. In addition to this decoction made from dried shoots is used as a remedy for lung infection. A part from human ailments this plant is also used to treat veterinary disorders. Leaves of this plant are given to horses to cure scurvy.

9.18.7 Toxicity

Use of this plant in excess amount can cause insomnia and anxiety.

9.19 *Swertia petiolata* D.Don

9.19.1 Systematic Study

Family	:	Gentianaceae
Synonym	:	*Swertia lahulensis* A. Kerner
English name	:	Tikta
Elevation	:	5000 m
Life form	:	Perennial herb
Conservation status	:	Threatened

9.19.2 Botanical Description

It is an erect and glabrous herb that attains average height of 89 cm. Shape of basal leaves is lanceolate; they are long-stalked and clasping with the stem at their base. Upper leaves are broad, lanceolate in shape, acute from margins and stalk less. Stem is erect and with 19–58 cm long. Inflorescence is spike and flowers are borne as clusters on terminal axis. Flowers vary in colour from bluish-white to purple or yellow with blue patches on them. Lobes of corolla are narrow to elliptic, length wise it is less than 1.3 cm having 1–2 basal nectaries. Fruit is in the form of capsule which is oblong to ellipsoidal in appearance. The seeds are several in number and reddish-brown in colour (Fig. 9.19a–b).

Fig. 9.19 (**a–b**) Field photograph of *Swertia petiolata*

9.19.3 Distribution

Afghanistan, Pakistan (Naran, Gilgit), Kashmir (Ratti Gali), Nepal, China. This plant is endemic to Himalayas.

9.19.4 Phenology

July–November

9.19.5 Ecology

The plant is commonly found in open alpine meadows, moist and wet places.

9.19.6 Medicinal and Edible Uses

The entire plant is highly medicinal. The most useful part of this plant is shoot, which is used to cure stomachache, vermifuge, febrifuge, appetizer, laxative and anthelmintic. Decoction of shoots is helpful in detoxifying liver and cure digestive system disorders. A paste made from crushing leaves is used as a poultice for maggot-infested ulcers. The powdered roots are useful in cholera. It is a bitter tonic and is administered for chronic fevers, anemia and general debility. It also helps to cure gastric problems. It has an anti-inflammatory function as well. It is considered to have a bitter taste and extremely cooling potency. It is used in the treatment of scleritis, inflammation of the liver and high blood pressure, inflammation of the stomach and liver. The plant yields a yellow dye, which is used for coloring fabrics.

9.19.7 Toxicity

Excessive intake of this plant can cause liver toxicity.

Chapter 10
A Pictorial Guide to Wetland Plants of Himalayas

Figures 10.1, 10.2, 10.3, 10.4, 10.5, 10.6, 10.7, 10.8, 10.9, 10.10, 10.11, 10.12, 10.13, 10.14, 10.15, 10.16, 10.17, 10.18, 10.19, 10.20, 10.21, 10.22, 10.23, 10.24, 10.25, 10.26, 10.27, 10.28, 10.29, 10.30, 10.31, 10.32, 10.33, 10.34, 10.35, 10.36, 10.37, 10.38, 10.39, 10.40, 10.41, 10.42, 10.43, 10.44, 10.45 and 10.46

Fig. 10.1 *Lemna minor*

M. A. Butt et al., *Wetland Plants*, https://doi.org/10.1007/978-3-030-69258-2_10

Fig. 10.2 *Nymphaea nouchali*

Fig. 10.3 *Ranunculus aquatilis*

Fig. 10.4 *Nymphea alba*

Fig. 10.5 *Potamegeton nodusus*

Fig. 10.6 *Marsilea quadrifolia*

Fig. 10.7 *Azolla pinnata*

Fig. 10.8 *Spirodela polyrhiza*

Fig. 10.9 *Alisma plantago aquatica*

Fig. 10.10 *Sagittaria trifolia*

Fig. 10.11 *Schoenoplectus lacustris*

Fig. 10.12 *Cyperus alternifolius*

Fig. 10.13 *Juncus articulatus*

Fig. 10.14 *Stellaria aquatica*

Fig. 10.15 *Ranunculus repens*

Fig. 10.16 *Eclipta prostata*

Fig. 10.17 *Eichhornia crassipes*

Fig. 10.18 *Veronica anagallis-aquatica*

Fig. 10.19 *Nasturtium officinale*

Fig. 10.20 *Saccharum spontaneum*

Fig. 10.21 *Oxalis debilis*

Fig. 10.22 *Asclepias curassavica*

Fig. 10.23 *Typha latifolia*

Fig. 10.24 *Canna× generalis*

Fig. 10.25 *Cardamine hirsuta*

Fig. 10.26 *Bacopa monerii*

Fig. 10.27 *Rorippa islandica*

Fig. 10.28 *Alternanthera sessilis*

Fig. 10.29 *Phyla nodiflora*

Fig. 10.30 *Eleocharis palustris*

Fig. 10.31 *Vallisneria spiralis*

Fig. 10.32 *Potamogeton perfoliatus*

Fig. 10.33 *Stuckenia pectinata*

Fig. 10.34 *Potamogeton lucens*

Fig. 10.35 *Rhodiola coccinea*

Fig. 10.36 *Swertia petiolata*

Fig. 10.37 *Leontopodium himalayanum*

Fig. 10.38 *Polygonum affine*

Fig. 10.39 *Ranunculus sceleratus*

Fig. 10.40 *Ranunculus muricatus*

Fig. 10.41 *Persicaria maculosa*

Fig. 10.42 *Acomastylis elata*

Fig. 10.43 *Cyathocline purpurea*

Fig. 10.44 *Polygonum plebium*

Fig. 10.45 *Portulacca oleracea*

Fig. 10.46 *Potentilla supina*

BM

The biodiversity of wetlands has great economic and aesthetic value and this diversity is helpful in maintaining overall environmental health. Most of the people around the world depend on aquatic resources to utilize them as food medicinal purposes and also as commercial purposes such as tourism and fishing. The Himalayan region has many wetlands of different sizes and shapes, which make it a distinctive ecosystem that fulfills an essential function in the overall water cycle of the basins. Wetland found in the Himalayan region is important as they provide shelter and food to the living organisms. Pasturelands on the banks of wetlands are used for grazing, and have also been known as the home of several rare endemic species of fauna and flora. The main purpose of this book is to explore the wetland flora of Himalayan region, to identify those plants which are beneficial to mankind and to aware people about the challenges faces by these wetland areas and the conservation strategies adopted for these wetland areas. This book not only helps with identification of wetland plants, but also helps in conservation and protection of most valuable resources.

M. A. Butt et al., *Wetland Plants*, https://doi.org/10.1007/978-3-030-69258-2

Index

M. A. Butt et al., *Wetland Plants*, https://doi.org/10.1007/978-3-030-69258-2